Global Modeling and Discontinuous Change

Global Modeling and Discontinuous Change

Presentations from a Conference of the World Growth Policy Group November 1983

Gerald R. Stairs
Editor

Duke University
Center for International Studies
Durham, North Carolina
1984

Occasional Paper Series
Number Fifteen

Library of Congress Catalog Card Number: 84-72483
ISBN: 0-916994-26-0

Cover illustration reproduced from *Funktion und ihre Darstellung in der Werbegrafik* by Anton Stankowski (New York: Hastings House, 1967)

This conference was created and arranged by the World Growth Policy Group following its earlier sector-by-sector analysis of the structure and substantive assumptions of some of the major global models. It was also the culmination of three intensive group seminars focussing on discontinuous change which were held in Fall 1983.

THE WORLD GROWTH POLICY GROUP

Joe W. Straley, Chairman

Richard N. L. Andrews, University of North Carolina
Moye Freymann, University of North Carolina
Norman Gustaveson, University of North Carolina
Michael Hoppe, University of North Carolina
Paul Jellinek, Robert Woods Johnson Foundation
Richard King, North Carolina State University
George C. Myers, Duke University
Scott Moreland, Research Triangle Institute
Richard C. Rowson, Duke University
Marion C. Salinger, Duke University
Robert Smythe, Carolina Resource Consultants
Gerald R. Stairs, California State University, Chico

CONTENTS

1

FROM PAST TO BEYOND

Gerald R. Stairs

Precis: Before this conference addresses the subject that brings us together, I will spend a moment or two, for those who were not involved in the nascence of the Group, to retrace the steps which brought us to this effort.

This project began about two years ago, with a group of faculty and development practitioners in the Triangle area of North Carolina joined together to reflect on the intriguing subject of global modeling. Since we were then about to celebrate, world-wide, the tenth anniversary of the World 3 Model and of Limits to Growth, a popular version of that study, it seemed to us that it would be useful to go back to the original model and examine its original data base. We undertook to follow that data base forward for the ten-year period and to find out how useful the model had been over that period. We are all cognizant with data bases: how inconsistent they are from year to year, and some of the other attendant problems.

Obviously it wasn't too long before the Group discovered that we would need to do more than simply examine the Global World 3 Model. Rather quickly we came to a philosophical confrontation with the dimensionality of the problem and found a need to examine the polar positions: the limits theories, the Cornucopian theories, and all of the range falling between. We spent a fair amount of time in discussion, in a series of reflective sessions, talking about the data base limitations and consequent problems. Some members of the Group were quite happy to consider the current versions of global modeling, while others talked about their constraints and suggested a need for further model development. We went on to look at an array of models: the multiple issues types, the large world models which have been developed, and a sampling of those now in operation or available for examination. We looked at regional models, commodity models, and some of the country models. We asked ourselves about the interrelation between these families of models: is there a network capability? who in fact is putting a network together to link regional, country, and commodity models to any current global modeling?

The Group achieved a good deal of perspective through participation in these varied discussions. One consistency became apparent in the early meetings: we had a subject around which we could gather a wide range of disciplines, and this kept the participant interest and interaction very high. In the early days of our meetings, there were a few colleagues (perhaps the more analytical of the Group) who entertained the notion that we ought to build a better world model; fortunately, we did not succumb to that notion. Greater wisdom prevailed, and thereafter we concentrated on an examination of the various dimensions of world models and on their utility.

About a year later we came to a summary point and prepared a panel presentation, based upon observations made during our group discussions for the 1982 international meeting of the Society for International Development (SID) in Baltimore. Our approach was to locate the field of global modeling on the learning curve after ten years of evolution. We suggested what uses could be made of world models, what limitations they present (or which one may encounter), and hazarded future estimates of their refinement and utility.

The SID international meeting provided something of a benchmark in our discussions. Response to the Baltimore presentation was positive; we were encouraged to pursue a leadership role in the field and our interdisciplinary approach was much applauded. The Group continued its interaction and has sought to sustain studies on the utility of foresight capability. Out of this experience came the idea of a conference on the implications of discontinuous change and what it means to modeling.

Another important realization came to us as we progressed in our thinking over the last two years: while we could gather the various disciplines around our investigations of model utility, we also needed to recognize that separate schools of inquiry have been developing over the past decade. The significant one (for our purposes) is the world modeling group. But there are other groups which talk about "world systems analysis" (e.g., World Systems Analysis, Hopkins and Wallerstein, Sage Publications, 1982). We recognize that these prognosticators (most of them are social scientists, political scientists, a few historians, and anthropologists) have spun off a set of texts and substantial studies explicating their theories. So there is a parallelism between those who are coming from the economics/resources/statistical side of the counter, who are using more analytical models, and those who are working directly from the social science (political/anthropological/historical) viewpoints. I believe the latter are identifying one kind of expanding economy in the world--assigning that role to the expanding multiple states. They speak about capital/labor relations a little differently perhaps from how the economists and resources people talk about the growth necessities; they give more weight to the political structures and refer to "core" and "periphery" areas in terms of nation-state development. A significant focus of their analysis has to do with those phenomena, differences, and measures of progress between states in their technological development which affect societal developmental phasing. I think they refer now and then in their more recent writings (as Toynbee did when he talked about the Industrial Revolution) to the fact that we are now in the "informatique" revolution, as the French critics named it--the era of "information technology"--or the "micromillenium," as others have characterized it. This is the era which our own Academy of Sciences has suggested might have more impact on humankind than did the Industrial Revolution.

It is obvious that a good deal of work has been done by the world systems analysts and perhaps we, from our side, have not paid as much attention to their approach as we should. It is now important for us to consider that there is indeed a recognizable body of such knowledge and opinion calling for interaction with our own discussions. We have taken some of their ideas into consideration from time to time when our members raised issues of political and social connectives. Therefore, I suggest to you that, as this conference looks beyond the current agenda, it might be useful to design a format for further work on discontinuities which would include world-class scholars who could bring these two schools of thought together. The world systems analysis groups, covering the political and social science arena, and the world modeling groups who deal more directly with current and prospective concerns for economic analysis and forecast, natural resources, population, and environmental quality, could complement each other.

Returning to the present conference theme, it is expected that our discussions will range widely. We will be pressed to resist the tendency to pursue issues of data inadequacy, model structures or assumptions, and related matters, as we attempt to deal with discontinuity. I am confident that our discussions will lead to a clearer understanding of modeling in relation to discontinuity and that, in itself, will have high utility to the holistic set of issues which interests us.

I am pleased to have the honor of presenting such a distinguished group of participants to the purpose of our meeting; on behalf of the local hosts I thank you for your willingness to join this conference and look forward with pleasure to your comments.

EDITOR'S _EX_ _POST_ INTRODUCTORY COMMENTS

The conference design called for a focus upon discussion of discontinuous change, a major issue in addressing global futures. Selection of participants was made on the assumption that a range of backgrounds and expertise was desirable, and that the theory of catastrophic event or major discontinuity on a global scale was not well developed. That we were successful in attracting to our discussion a wide range of viewpoints and backgrounds is evident from the following papers. Those who have stood on the firing line of decision-making have shared views with the theoretician and the academic.

The conference presented, in the aggregate, an excellent exchange--one that the Group felt should be preserved and shared in print. Thus we have collected in the following papers the primary comment from major presentations. With an eye to providing continuity of viewpoint by author, we have brought the two days of discussion and contributions into single-paper format, by author. While this does forego the dynamics of interaction, the editorial decision was made in order to allow more concise transfer of comment in the printed form.

I trust that the following papers will be of service to the reader interested in issues of taxonomy and concept in viewing the role and predictability of global discontinuous change.

TOWARD A TAXONOMY AND UNDERSTANDING OF DISCONTINUOUS CHANGE

Robert B. Smythe

Our study of the implications of discontinuous change for the use of models began with the preparation by certain members of the Group of specific briefings for mutual discussion at a series of weekly seminars. At times these were in the form of presentations on national planning, economic forecasting, technical issues, and a variety of other connected views--some from our own members and some by invited practitioners and scholars. Growing from these meetings, a lead commentary was prepared to serve as the basis for further discussion at the Quail Roost conference in November, 1983. Contributions by Robert L. Hoffman, Richard A. King, R. Scott Moreland, Timothy J. Lomperis, and Jose Carballo-Vega have gone into the discussion section which follows, and which serves as a foundation for the position presentation I now bring to this conference.

I

The Group offers a number of definitions of discontinuities:

> "Any kind of sudden change in the state of the world."

> "Any change in a real world parameter which had not been predicted by the model which purported to include it."

> "A significant departure from an established trend."

Discontinuous change, whether abstractly or operationally defined, presents difficulties over time for persons attempting to model global or regional systems (political, economic, ecological) if such models are intended to provide a predictive or foresight capacity for future real world events. Existing models have been criticized both for their failure to provide for the inclusion of discontinuous events as variables in the system, and for their failure to deal realistically with discontinuities which may be produced by the performance of such models. Because of these acknowledged weaknesses, the direct use of such models in government and private-sector planning and decision-making has been rather limited.

Discontinuities may be described or classified according to many different criteria. They may be foreseeable or not, manipulable or not. Their occurrence may be focused or diffuse, instantaneous or progressive. A discontinuity may have single or multiple causes, and in turn may spawn future discontinuities through multiplier effects. The effects of a discontinuity may be short or very long. An example might be the capture of the lower Mississippi River by the Atchafalaya River--an event which some experts consider likely. Such a change would have multiple causes, a single precipitating event (failure of the man-made structures which have prevented it to date), and, if not quickly reversed, drastic long-term consequences for the cities of Baton Rouge and New Orleans and the larger economic and political systems dependent on the river's commercial traffic.

Examples of Discontinuities

Nuclear war	Regional conflict	Military coup
Pandemic	Famine	Crop failure
Climate change	Drought	Flood
Depression	Recession	Market cycles
Global urbanization		Regional emigration
Disease eradication		Improved contraceptive
New energy source	New food source	New crop variety
Global arms treaty	Bilateral treaty	National election
New technology		New scientific discovery

There is an implicit classification in the above list of examples: from top to bottom--destructive to beneficial; from left to right--global to local (or major to minor). But the scales selected for this classification are not operational ones, i.e., they do not help achieve some defined purpose or goal.

Suppose we define, as the objective, a classification of discontinuities which will aid the modeler of geo-political systems at the nation-state level or larger to identify (and attempt to incorporate or predict) those discontinuous events which should be of greatest significance to national policy analysts and decision makers over the next ten to twenty years. We have then narrowed and defined the problem so that some meaningful criteria for a classification may emerge. We would argue that for this purpose there are three principal scales or axes by which discontinuities should be ranked or, if possible, simultaneously arrayed:

A. Predictability (Levels are given to suggest a range.)

1. Foreseeable, but more data required to become predictable (commodity price cycles, etc.)

3. Partly foreseeable, timing and consequences uncertain (cure for cancer, regional resource conflicts, etc.)

4. Not foreseeable (discoveries, natural disasters, etc.)

**Modelers have attempted to deal with type 1, and to some extent (on a "what if" basis) with type 2. Policy analysts have to give some consideration to all three types.

B. Manipulability (A range is again suggested by three levels.)

1. Reversible or preventable by actions available to the decision makers (some military conflicts and natural disasters, many technological changes)

2. Subject to control of timing, intensity, duration (many economic cycles, most scientific and technological changes, some demographic phenomena)

3. Irreversible or nonmanipulable (some natural disasters, climatic changes, discoveries, mutations, biological or physical limitations on resource availability)

**This criterion is of vital interest to government officials, corporate executives, and investors, but has not been explicitly recognized in most global and regional models.

C. Significance of Effects

This criterion is a composite one which is difficult to define apart from a specific model; it should, however, allow a ranking of the effects of discontinuities based on their expected scope (geographic or demographic), their intensity (disruption/enhancement of system stability), and their persistence (time required for recovery/assimilation). Thus a scale of significance might encompass the following:

1. National/international, overwhelming, long-term effects (world war, deforestation, smallpox eradication, etc.)

2. Regional, intense, medium-term effects (earthquake, drought, election, assassination, treaty, etc.)

3. Local, mild, short-term effects (flood, oil spill, crop surplus, development project, etc.)

It might be argued that this "effects criterion" is the only one of interest to decision makers; however, unless a potential discontinuous event is reasonably foreseeable, as well as subject to some control or modification, it is likely to be given scant attention--no matter how significant its effects might be. Thus the combined relationships of the rankings of a potential discontinuity on all three of the above axes should be of interest to the system modeler.

If one represents these relationships via a three-dimensional plot, in which the intersection of the three axes is the point of maximum value on each scale, then those events which are located closest to this intersection should receive the highest attention of both modeler and decision maker. Such a plot is suggested in the diagram below.

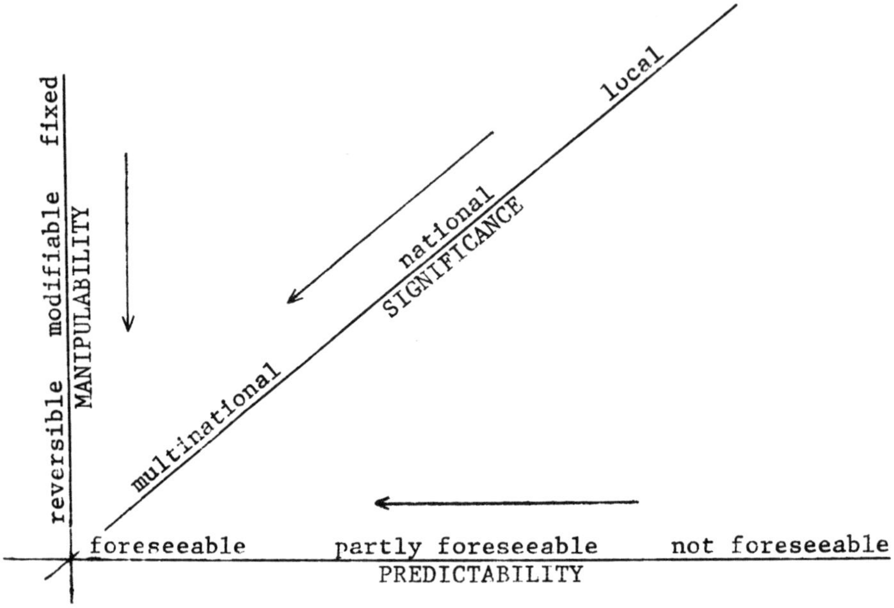

Figure 1

II

The purpose of my position paper, which derives from and extends the above backgrounding, is entirely heuristic: to devise some framework which may have been provoked by what we have read, and not to attempt to answer all questions. Let me simply outline the setting for our reflections.

The Group has talked for some time, as is the intent of this conference, about the nature of discontinuities. We originally used the phrase "seismic." Some people felt that was too geological. The idea of "discontinuity" can be defined in different ways--we have not tried to give you one definition. I like a modest one: that it is any change in the real world which was not predicted by a model that was intended to provide foresight. But the element of surprise is supposed to be common to any definition--an unanticipated change. I think that implicit in any discussion of discontinuity is not the event itself, but the effect it has on paradigm shift--if you like--in the academic world, or on political change, or on the fairly long-term, if not permanent, environmental/ecological shift which may result.

The problem that has confronted us, and led to our coming together as an interested group, is the fact that existing models do not deal well with discontinuities--neither incorporating them into models nor predicting such events on the basis of whatever variables are in the model.

We found that the concept of discontinuous change is a basic problem to modelers, and to anyone attempting to devise a means of systematically foreseeing future events or providing for such capability. This is a somewhat less ambitious way of describing unforeseen changes to people who make decisions about economic and political systems. Now there is a tendency for us to imply that we are talking exclusively about heads of state; we are not necessarily doing that. We are talking about people who are involved in government or are managers of resources. These are probably heads of political organizations to be sure, but we are also talking about chief executive officers of major corporations whose decisions are based on global forecasts of resources and economic events.

Now I suppose this is the time to reveal my own bias. I spent some seven years with the Council on Environmental Quality and I struggled to apply what we hoped were concepts fairly well understood in technical and scientific circles involved in the formation of policy, and hence to the execution of events and actions by appropriate administrative agencies. That is not the only reason for my interest, but my remarks may betray that bias. It is obvious to me that, because of weaknesses and failures in predictions, direct use of models (global or otherwise) in decision-making has been rather limited. There is in fact a study by Dr. Don McGray, of the East-West Center in Hawaii, and his colleagues, which reviews most of the world models and regional models as of about two years ago. These researchers asked this very question: to what extent had models been used directly to affect political decisions and large-scale economic decisions? I found very little evidence that there is any direct effect, but I found much evidence of indirect effects. That may be the only way effects will continue to occur, but nevertheless that discovery causes us to be concerned, and the question then is: can we determine something about the nature and classification of discontinuities which will allow modelers to deal better with them, and hence be able to communicate better with people in either government or sector decision-making roles?

We can make a few points: (1) that discontinuities can be foreseeable to some extent; (2) that they also can be totally unforeseeable; (3) to some extent they can be manipulable or reversible or not; (4) they can have locally-intense effects which are not significant on a larger geographic scale; or (5) they can have slighter but very long-term and sometimes global consequences. In some auxiliary comments at our conference, Chancellor John Caldwell has pointed out that the concept of world urbanization is, in some respects, a discontinuity that depends on your time scale, your geographic scale, etc.

The examples of discontinuities in the discussion paper are both implicit and intuitive. What we are really saying is that you can't classify or develop a taxonomy for discontinuities unless you have an operational goal in mind. If, for the purposes of this discussion, you take as the objective the classifying of discontinuities to aid someone who is modeling geo-political systems at (roughly) the nation-state level or larger, then there are certain ways of sorting and classifying discontinuities that may be more important than others. For the sake of argument, we would suggest that if you accept that rough objective, then three axes--three scales--seem to us more important than any others, although they are not necessarily the only ones. One scale shows the extent to which discontinuity is or could be predicted. Many events, both good and bad, used to be totally unforseeable. However, we are beginning to learn more about earthquakes, for example, and are now able to predict some seismic events. We are also able to say some things about climate which might make predictions more accurate.

For those who are making decisions about resource utilization, or political stability, reversibility and manipulability are important. If an event can be changed by having been foreseen, then you find decision makers becoming very interested in the potential discontinuity and asking, "How can I get involved in preventing it if it's bad, in enhancing it if it's good, and take credit for it immediately?" One can see the utility in forecasting for the political sector! It is also fundamental if one is interested in commodity futures speculation, or in making major investments in the world resources market. If you can predict the discontinuity early, and then manipulate, you are generally pretty well off. (You may also suffer, if your model is inadequate or you assume too much manipulability.)

Finally I have suggested a third index scale which I leave to you to refine, if you like, and give a better name than "significance." That scale has elements which can be really defined only when one asks, "Significant in what framework? In regard to what modeling? In regard to what geographic area? In regard to what time scale?" The key here is not significance of the discontinuity, but significance of effects. You know that an earthquake, or a coup in Central America, or an invasion by a North American power are discontinuities--but the question is, what is the scope and intensity and persistence of the effect? Hence the scope of effects, intensity of effects, and persistence of effects are all elements which can be collapsed into a scale of significance so as to leave one impression on the persons looking at the output of this model.

We drew those axes together on a three-dimensional scale [Fig. 1]. The three lines are three axes, perpendicular to each other so that one of them--the significance one--comes out of the paper. If one can define the system in which one is interested, and define the range of potential discontinuities (at least the ones that are imaginable), the question to the modeler then becomes: which of these potential discontinuities merits his time? The hypothesis is that those discontinuities which are closer to the intersection of the three axes will be of more importance or utility in the model than those which are further away. Obviously, if something is totally foreseeable it is no longer a discontinuity, and that presents a certain problem: some of that aspect of surprise which we generally attribute to discontinuity may be lost.

But whether the axes intercept or not, whether there should be a fourth or fifth dimension or others, is perhaps not important. What we are trying to achieve is a clarification of our objectives in trying to define discontinuities and to relate the kind of modeling that has been done in the past to the kind that might be done in the future. I would like to see the conference participants improve the design we have put before you.

3

ISSUES, ADDRESS, AND SOLUTIONS IN DISCONTINUOUS CHANGE

Barry B. Hughes

INTRODUCTION

When Dr. Stairs introduced the session tonight a realization came to me which I am almost reluctant to admit: he referred to the tenth anniversary of the Limits to Growth study and this is also the tenth anniversary of my own involvement in world modeling. I started with the Mesarovic-Pestel group back in 1973. That does not necessarily establish my credentials as a modeler--in fact the more you work in the field the more you realize your own limitations. But perhaps it helps me to establish some credentials in terms of joining the list of those who critique models and who identify the strengths and weaknesses of the modeling process.

For a number of reasons, about four years ago I decided to change directions and began developing a model which I call International Futures, or IFs. This model is geared toward use in classrooms--specifically in college classrooms--and was so designated because of the effort of the model to answer or to address international futures kinds of questions, many of which obviously involve discontinuities, as well as to take inputs (if-statements) suggested by students or other users and try (in some sense) to respond reasonably.

There were a number of reasons, a number of bases, for my decision to move away from more traditional modeling with a policy analysis or forecasting orientation. Some of our colleagues have reviewed a number of these models in terms of their limitations, for use in the policy process. Among these limitations, obviously, were discontinuities themselves--our inability to predict or forecast them in a way that ensures them to be useful instruments for many policy makers. Some of the reasons don't center around discontinuities: they center around our theory and our ability to model continuous change and continuous series of events. We have as many problems there (we may as well admit) as we do when handling discontinuities.

Dr. Stairs spoke earlier of the possibility that this Group might bring together people who have been involved in world modeling projects and people from the world systems school or systems perspectives. I am reminded thereby of the tremendous differences in theory or views about the world with respect to continuous change and continuous processes. Just a very simple example: let us take a global model and try to deal in some way with the impact of foreign aid transfers, transfers from one region of the world to another. Most of the world models which have been developed incorporate positive linkages into foreign aid transfers: transfers increase investment availability or increase foreign exchange. This is obviously traditional in most two-gap economic models. The world systems school, looking at that same transfer, might well question the positive impacts, citing distortions which the aid imposes on recipient economies and the political control which accompanies that aid connection. It is therefore apparent that even the continuous kinds of relationships divide schools, divide perspectives, to such a degree that it is very difficult to identify common theories underlying all forecasting models. I won't even touch on the data question because that topic was specifically excluded from this agenda--but we all know the difficulties presented there.

Given our problems in dealing with discontinuities, our inability to agree on theory, and our data problems, any long-range forecast produced by a world model clearly has a low-probability future. This leads me to what I consider to be a more proper use of models--and here I agree with many things that John Sterman said. I think that models can help us understand systems: what we know and what we don't know about them, where we agree and where we don't agree concerning them. Models can help us organize our own thoughts as modelers, as students of global systems, and they can help us communicate bits of knowledge and bits of uncertainties to others. That's also true in dealing with policy makers. I think it is the primary contribution which modelers can make and that is what has impelled me in the direction of the development of an educational tool. I think modeling can be incredibly useful as an educational device. In the Mesarovic-Pestel project, to the degree that we were able to work with policy makers, I found over and over again that what we really contributed was a greater understanding about global systems, rather than specific policy recommendations. In all honesty I have to say that we didn't have a great deal of contact with people making policy but more with people working in advisory roles to policy makers.

Now let me turn more directly to the issue of discontinuities. One of the things I would suggest as being useful is to emphasize a distinction which was implicit in the document that Dr. Smythe discussed. It is the distinction between forecasting discontinuities--that is, building models which tell us some discontinuity is likely to occur in the global system--and realistically modeling the impact of various discontinuities (posing the "what if" kinds of questions). I think that's a very important distinction because the two types of approaches to discontinuity vary considerably in difficulties encountered. The first--forecasting discontinuities--is an area in which global models, or any other kind of model, will always be confronted with tremendous problems. Approaches to forecasting of discontinuities using probabilities, using thresholds--for instance, catastrophe theory--are useful and should be pursued, but I am not very optimistic about their doing any more than what John Sterman said: they can give us only the sense that a certain kind of discontinuity might occur and that there is a fairly wide probability range associated with it. I would be reluctant to go to a policy maker with a model and say, "Here is a probable discontinuity, here is a probable change in the system."

The second portion of the issue of discontinuities is modeling their impact. Here we can reasonably expect models to improve considerably. Let me just give one example from the International Futures model I developed. This model was taken up for study by a group at the National Center for Atmospheric Research. Although the model normally runs from 1980 forward, they initialized it with 1970 data, and studied the agricultural issues of the 1970's with it--particularly problems in global agriculture between 1972 and 1974. The first thing that happened is obvious--that model did not produce any discontinuities in 1972 and 1974 in the agriculture system on a global basis, no matter how well it was initialized with 1970 data. There is no climate system in the model; even if a climate system were included in the model, it would not have forecast the weather problems we had at that time. So, on the first issue, the model fails completely; I am not particularly upset about that. The model was not constructed to forecast agricultural discontinuities and I think that no matter how much effort might have channeled into the model, it would not have dealt with the issue adequately.

The Atmospheric Research people took the model and artificially introduced the impacts of the weather problems in 1972-1974 on yields in various regions of the world--lower yields in Africa, representing the Sahel impact, lower yields in Asia and the Soviet Union--in an effort to see what the secondary responses of the model would be. Here, too, the model failed in many respects. It indicated the kind of price run-ups that should result; it indicated some of the other secondary impacts; but, in general, it over-reacted considerably. Within a year or so the model was producing agricultural surpluses that were considerably greater than the real world agricultural surpluses which in fact followed the same period. Obviously the model generally failed in that regard. Here is a case in which one could go back to the model and fairly easily improve its ability to deal with that specific kind of discontinuity. If a model is constructed to answer that kind of question--which it may not be--such work should be done.

We have all been fairly hard on models, detailing their limitations. No doubt such a stance is justifiable, but I would like to end by saying that after ten years I am still working on models and I am very positive about the things models can do--particularly in the communication/education area. We have to realize (and we do, in this group at least) that when decisions are being made or when discontinuities are being considered, the mental models, which are the possible substitutes for computer models, have at least equally great limitations.

POSITIONS ON DISCONTINUITY

A. The Issue

The issue needs to be subdivided into two parts:

A. Global models do not forecast discontinuities or shocks well, primarily because modeling is basically an extrapolative exercise. Approaches to improving model behavior through random introduction of shocks and identification of thresholds where behavior might change (à la catastrophe theory) are sometimes useful. However, expectations here should not be great.

B. Global models often do not respond correctly or reasonably when "shocked" endogenously or exogenously. In many respects this is the more serious failure from a policy analysis viewpoint. It is also more correctable. These remarks will therefore focus on this aspect in the future.

B. Evidence of Failure

1. My own International Futures (IFs) model can illustrate both types of problems with respect to the agricultural sector. First, it makes no attempt to suggest discontinuities in food production resulting from regular cycles or irregular climate/weather changes. Second (and disconcerting), an attempt to introduce the yield losses of the early 1970s results in inappropriate model response. Specifically, the model over-reacts in terms of subsequently increased production after one or two years. The major reason is absence of investment response gestation periods and of some forward-looking decision capability.

2. The problems affect all global models. For instance, WIM had an identical problem with responses to oil embargoes.

C. Analysis of the Problems

1. Models are extrapolative. Identifying turning points, thresholds, etc., is at or beyond the limits of their capabilities--at least, stating _when_ they _might_ occur is; but identifying a high _probability_ that they

will occur is not. If the first problem (in the issue identification) is considered more important, an approach using probabilities of discontinuities (themselves continuous) would be appropriate.

2. Global models are geared toward the long term and toward such issues as availability of food and resources, or the shape of demographic and energy transitions. Therefore they purposely leave out such features as delays, distinctions between short- and long-term elasticity, and gestation periods, when these are not seen as necessary in sketching appropriate long-term behavior of the system.

3. Much of discontinuity representation requires representation of political, social, or psychological factors. For example, during the oil shock period of 1979-80, 1979 global oil production was five percent above that of 1978 while consumption went down and stocks reached very high levels. The reason was hoarding--when prices went up, demand went up and supply went down, the opposite of results dictated by the economic theory normally imbedded in models. Similarly, analyses of income distributions show them to be remarkably stable, and models which represent distributions at all, represent them as stable. They change primarily after revolutions and significant social change. No global model has attempted to endogenize the social and political results of major income distribution change, as important as they are.

D. Solutions/Recommendations

1. Identify specifically the short-term issues or discontinuities of importance, e.g., crop failure, oil shock, revolution, war. For instance, the Pentagon already does this as part of its long-term planning.

2. Consciously build models to test the impact of such discontinuities. Research the delays, constraints, bottlenecks, and other factors which make short-term behavior significantly different from long-term behavior.

3. Introduce greater numbers of political and social variables, and relationships specifying how they

interact with economic and ecological mechanisms of response to discontinuities.

FINAL DISCUSSION

I would like to follow through the points in my original position paper and actually use it to organize these further thoughts. I want to touch on four basic subject areas, saying a little about the issue of discontinuity itself and how to conceptualize it, and a bit on the evidence of our failure to deal with it in modeling. We will now turn to a brief analysis of the problems, reasons why we fail to deal with discontinuity, and finally devote a couple of minutes to solutions and recommendations.

We should underline again the distinction between forecasting discontinuities and yielding the secondary and tertiary consequences of discontinuities once they are introduced through "what-if" analysis. With respect to the first: global models don't do well with discontinuities or shocks--or whatever you want to call them--in large part because modeling is an extrapolative kind of exercise. Models take a certain historical path and follow it forward in time. Models, as a rule, are just not designed to do much more than is dictated by extra-polite, smooth behavior. To be sure, there are a variety of suggested methods of introducing capabilities for predicting shocks or discontinuities; none of them, in my mind, have been terribly successful.

Many of you, I know, are political scientists, with some exposure to International Simulation (INS) and even to the computerized version of that simulation by Stuart Bremer. This was one of the few world models which really attempted to deal with the issue of domestic conflict and the possibility of domestic revolution. It relied on a random element. Certain factors in the domestic environment led to a higher probability of change in government, peaceful or non-peaceful, and a randomness introduced periodic non-peaceful change in government. But that isn't necessarily a very happy approach for most of us. We certainly wouldn't want to go to someone with a model and say it randomly depicted a discontinuity at this point and that therefore we should be careful. We might want to advise that "this model shows a high probability of a significant problem at this point in time." That would be a treatment of possible discontinuities in continuous terms. Once we start talking about probabilities, we're talking in terms of continuous variables again and are back on easier ground.

Many of us have had experience demonstrating the inadequacy of our models to forecast discontinuous change. I built my first energy model in the summer of 1973, which was not a particularly auspicious time for building energy models. In defense, I should say that at that time those of us who were working in this field were already seeing the turn-around in prices; real oil prices had been going down globally throughout the 1960's but did start going up in the 1970's. We did observe the forces that were creating a tightness in the world oil market. For instance, the U.S. was gradually eliminating its oil import quotas. In fact, we eliminated those quotas completely in early 1973, throwing ourselves into the world market and into competition with other supplicants for Middle Eastern oil supplies. So, many of us who were modeling energy did see the tightening and many of us suggested that prices were likely to continue on an upward trend well into the 1970's. But none of us forecast a four-fold price increase and, in retrospect, I still don't believe it could have been done.

The second approach to discontinuities in models is to introduce them on an if-then basis and see how the models respond. Here I'm much more optimistic about the potential. Models ought to be able to do this if they are structured properly. I think that this is the more useful area for us in terms of policy analysis. We may well be happy (on a policy basis) to introduce our shocks, to introduce our discontinuities, and have the models tell us how the world is likely to respond. That kind of information is extremely useful.

I should address my remarks more directly to this second kind of treatment of discontinuity than to the first. There is no need to repeat comments on evidences of failure. I talked last night about how my own International Futures (IFs) model often fails on both accounts--both in predicting discontinuities and in adequately responding to them. One of the reasons that it does not respond adequately to sudden introduction of yield restriction, simulating a climate or weather problem, is that there is a very short gestation period for agricultural capital. Most world models, including IFs, compute in one-year time steps: they do not have any kind of delay with respect to capital formation. A decision is made to invest and the capital comes onstream in the next time step. As a result, when a shock is introduced into the system, yields drop, prices rise quite significantly, the system responds with an infusion of capital into agriculture, and in the next time step there is additional food production. If weather changes so that you eliminate the weather problem at the same time that you have increased agricultural investment, you not only have additional food, you have way too much. It is a very simple thing to explain; there is no mystery about it.

I might call the last section of my remarks "back to basics" because I am going to say some things that are fairly fundamental but bear repeating. First, models are extrapolative. That means that we take a pattern of past behavior and we extend it into the future--whether that past behavior is a trend, or a set of cycles, or the pattern of interactions among multiple variables. To the degree that we deal with cycles and extend cycles into the future we can suggest turning points; the work that Dr. Sterman and his group are doing suggests that you can forecast turning points. But you can not do any more than extend into the future the pattern of cycles with the historic periodicity or timing of cycles. You cannot normally suggest that there is going to be a significant change in that cyclical behavior. The only real discontinuity in the cyclical behavior shown us (by Dr. Sterman) is the change in inflation rate. That figure indicated a set of cycles in which inflation rose and fell--I think about three cycles--and then reached a point where inflation did not come back down. The way in which that sort of thing is introduced into a model is different from the continuation of the cycles. I assume that was the point at which a permanent deficit took effect--accomplished by the introduction of an exogenous factor into the model. In short, the continuous cycles were endogenous but the one real discontinuity was not.

The approach here, in introducing what some might think of as discontinuity, is really to make it continuous; that is what we do with cycles. Another way of making discontinuities continuous is to try to quantify the variables in a different sort of way. As I suggested before, identifying probabilities as opposed to a threshold at which something is bound to occur--stating that the probability of an event moves from .4 to .5 to .6--is a way of creating a discontinuity.

Another example can be found in dealing with international or domestic conflict or cooperation. Instead of talking about these as events, we can talk about levels of international conflict in an event, using an approach called "events data analysis." Some of the political scientists in this group are familiar with approaches for classifying interactions among states in conflictual and cooperative terms on scales, ranging from highly conflictual to highly cooperative. There is a group in Berlin working on a model called "GLOBUS." They are one of the few groups in the world attempting to forecast political events; in particular they are looking at the possibility of international conflict. But rather than trying to forecast an outbreak of a conflict, either regionally or globally, what they have done is to make continuous the variable of conflict, to develop a scale (high conflict to low conflict), and to forecast the changes in that scale with response to changes in the international and national economic and political environments. Again, they have made _continuous_ what we've often thought of as _discontinuous_. That is one small element of the "back-to-basics" approach. Because models are extrapolative, forecasting discontinuities involves, for the most part, making them continuous--to the degree we can do this.

The second part of the "back-to-basics" approach is the recognition that global models are geared toward the long term. Obviously we have to remember that models are simplifications of a reference system--of a reality that we are trying to identify in simplified structural form. We simplify according to the purposes of the model. Most global models--the ones that you people have been looking at, whether _Limits to Growth_, the Mesarovic-Pestel model, Bariloche, or whatever--

have been concerned with long-term issues. They have dealt with long-term population trends, global demographic transition, long-term energy trends and transitions, and long-term food availability. Insofar as their basic concepts are concerned, they have not necessarily been engaged with what happens if you introduce a short-term weather fluctuation. As a result these models do not deal well with such an issue.

What the global models have omitted in their long-term orientation are factors like delays--in many cases short- versus long-term elasticities. Why introduce a short-term elasticity for energy demand response to price changes if you are really interested in the long-term equilibrium? Constraints of various types would slow down response mechanisms. Such things have been left out of these world models, but that does not mean that they couldn't be put in if world modelers wanted to deal with these issues.

The third and final "back-to-basics" comment on models and their dealing with discontinuity builds a little on what Lena Lupica said last night about the political and social elements which have been omitted. Many of those love the nature of continuous change. I have been studying comparative economic and policy responses to the two oil shocks in 1973-74 and 1979-80. There is a great deal that cannot be adequately explained in those shocks by more traditional kinds of economic models. For instance, in the 1979 oil shock production of oil increased globally by 5% above that of 1978, but consumption of oil was dropping. This makes one begin to wonder how much shortage there really was.

In reality, global oil stocks at the end of 1979 were significantly above global oil stocks at the beginning of 1979, yet prices went up dramatically during that period. Of course what happened was, in large part, hoarding. As soon as Iranian workers struck in late 1978 and cut off supplies of oil, people began to worry about maintaining their own consumption and began to purchase oil. This happened at both the consumer level and at the company (distributor) level. People were hoarding the oil, taking it off the market, and the price was being driven upward. Higher prices became a self-fulfilling prophecy and the results contradicted the traditional economic logic. Prices went up; therefore supplies were supposed to increase and demand was supposed to decrease. What happened was that prices went up, demand increased, and supply decreased. You can build that sort of thing into a model, but you have to recognize the kind of political and psychological factors involved. And they are often discontinuous.

Another example from the oil shock periods is the differential impact of the two oil shocks on Japan and the U.S. in 1973 and in 1979. In 1973 Japan had extremely high nominal interest rates. It went into the larger of its two recessions; the 1974-75 recession for Japan was significantly deeper than the 1980-81 recession. In contrast, U.S. interest rates were much higher in the second period and our recession was deeper in the second period. But the interesting thing is that the inflation rate in Japan was higher in the first recession, and inflation in the United States was considerably faster in the second recession. These are elements that do not all fit together in terms of traditional economic theory.

I would argue that the major difference between the two periods and the two countries was the treatment of oil prices. During the first oil shock, most of the higher global oil prices were passed on to consumers in Japan, but the U.S. controlled oil prices then. In the second oil shock, the United States passed on both the first and the second oil shock price increases, by decontrolling prices. These politically determined the pricing policies with interest rates, with money supply policy, and affected the larger economic response of the system. That is a policy decision which would have to be introduced into the model if it were to treat adequately economic issues in those periods. And it can be introduced. It is a question of whether these are the issues with which you wish to work.

Now I will submit a final recommendation with respect to what we can do about discontinuity. We have to focus on identifying the discontinuities of particular importance to us. The list of discontinuities we considered in this conference is very useful, but no world model will ever deal with such a broad set. What a world model team might be expected to do is to select a handful of those discontinuities and deal with them successfully. We must consciously build our models to test the impact of specific discontinuities. To introduce such treatment we will have to research the delays, the constraints, the bottlenecks, and the other factors which could significantly affect the impact of discontinuity on the broader system.

I would say, in conclusion, that it is time for us to introduce more political, social and psychological factors into the models--particularly if we are going to deal with the impact of discontinuities--because these interact strongly with the way in which shocks affect the broader system.

4

DISCONTINUOUS ECONOMIC CHANGE

Richard H. Day

The world, the empirical world, is quite discontinuous; it exhibits complicated dynamics. Many people find this disheartening, so they refuse to see it and insist, instead, that the world is very stable. It is appropriate to remind ourselves of what some of the empirical facts are around us and to see whether or not we have to alter the way we think about the real world.

I. Complicated Dynamics

Economists have long recognized the existence of cycles. The latter have always caused a certain amount of difficulty because their turning points and their amplitudes have always been hard to predict. A prevailing explanation for this is one that Dr. Sterman gave: putting random inputs into a stable system so that it generates cycles of various kinds; these will be perpetuated as long as the shocks continue to take place. New research, however, which I will review briefly below, shows that a complicated pattern of cycles can be generated with no intrusion of random shocks at all; their seemingly stochastic behavior can be intrinsic to the structure of the relationships in the system itself--even if that system is entirely deterministic. This means that the unpredictable reversal of turning points or change in amplitudes are things that we might be able to explain.

That leads to a sort of paradox: we can't _predict_ but we may be able to _explain_. If so, can we do anything about policy? Doesn't unpredictability destroy our role as scientists? I see it this way: I think anything that we can understand, we have a chance of controlling. And if we can explain unpredictability in our minds, there is the possibility that we can think our way out of some of the surprises and avoid them.

Certain economists often argue that we judge models and theories by how well they predict: if they can't predict then they are supposedly no good. My feeling is that if there is anything at which we as economists are bad, it's predicting--but if we can actually explain _why_ we can't predict, then we are back in theoretical business.

Rapid change is something that many people find difficult to appreciate; it's really a matter of time perspective. When I claim that the world is unstable, many people say, "Look out the window: is the world unstable?" If an earthquake hits right then, they might change their minds. But generally things aren't moving too violently out there and that seems to be proof enough. However, when you look at things over a longer time perspective, you can see change which is not otherwise discernable. If you look at population change over a short period of time it seems to be changing gradually. If you look at it over the lifetime of the earth, which is something like eight to ten billion years, then what you get is a vertical spike. All the dynamics of human history are encompassed within the thickness of the line that makes up that spike--almost all of it. What you see is an explosion. Obviously what we need to do is to take a little longer perspective sometimes, maybe not always so long, but longer than we often take.

Rapid decline occurs also. Civilizations do end and some of them end rather quickly. The one in Taotihuacan, south of Mexico City, apparently declined in approximately 75 years. Now that's a very long time for an economist's mentality, but for an archaeologist's mentality it is regarded as a rather sudden event because it came after some eight to ten centuries of flourishing growth. Again, it's a matter of time perspective.

What we see in all of this is a fairly rapid development. We can see overlapping waves of change as rapid growth occurs, rapid decline takes place, and then something else enters and begins growing rapidly. A good deal of purely conventional economic reasoning can explain part of this kind of development: as resources are used up, a switch has to take place. But switching from one technology to another doesn't capture the essence of what's going on--because a part of what's going on is the emergence of new things and that can take place fairly suddenly. The invention of iron-making may have taken place gradually, in a certain sense, and the ability to make iron existed probably for many centuries before it was actually an important technology. But eventually iron-making went through a very rapid expansion. Economic conditions were favorable; we got an overlapping wave as the iron age replaced the bronze age.

This kind of phenomenon is occurring at a faster and faster pace. There is an accelerating appearance of new things and an accelerating destruction of old things. We are just witnessing the last vestige of paleolithic culture as it disappears altogether from the face of the earth. It persisted for a long time, as long as civilization has existed--until now. On the other hand, many individual technologies which have been used by various civilizations have disappeared rapidly; some are now in the process of disappearing. Looking about one can see that the world is in a state of very rapid change and that new things are continually coming into existence. It seems to me that this really is the essence of discontinuity: the emergence and rapid adoption of things that are essentially new; the decline and eventual abandonment of things that are old.

In economics, of course, the great theorist of this process was Schumpeter, who argued that the entrepreneur played a special role in the process of economic development by creating new combinations. To Schumpeter, the entrepreneur was a creative force.

This creative force is not unlike Prometheus, who handed fire to man in a unique creative event which precipitated civilization that grew and flourished ever after. Prometheus is out there functioning all the time, and in some modest way, if I may say so, he may even be functioning in each one of us. The fact is that new things are being cooked up by some of us all the time. When the right time arrives these new ideas become either catalysts or the focal points for change, which then takes place in an almost explosive manner.

The central issue that has to be contested by intellectuals who build models and theories is: can these processes of sudden change (given that we recognize their empirical relevance) best be characterized by equilibrium models, by stable, dynamic systems, or should we be devoting a lot of attention to unstable systems and disequilibrium processes? That is the question I really want to focus on next.

I would like to close this section first, however, by responding to the remarks made earlier in this conference about the need for models to be relevant for policy makers. That point has been made frequently by policy makers, and of course by modelers as well, and it is well taken. It is important to try to be relevant. I'd like to make another plea, however, which is that policy makers should try to be relevant to models!

Policy makers often have certain questions in mind. They want answers to those questions from the modeler. But one of the uses of a model is to help articulate or identify what some of the right questions are. Sometimes the single most important input that a modeler has to make is the posing of different questions. The policy maker may be looking at a given set of questions which he thinks are very important because they spring from the political reality of the time. The modeler may come along and say, "You think that's an important question? Here is another question that's far more important and we must direct more attention to it." Policy makers can (or should) exploit the modelers more fully--asking them what <u>they</u> think are the important questions, and what <u>they</u> have to contribute from their unique perspectives.

Turning back to the modelers, the fact is that most modelers, most theorists, <u>do</u> think of the world as stable and they <u>do</u> use continuous and stable models. I think if the modeler is going to be relevant to the facts, it's time to give more attention to unstable models--ones that exhibit instability and discontinuity.

II. Dynamical Systems Theory

In this century techniques have been developed that help us model and analyze complicated dynamics. One broad field of technique can be called "dynamical systems theory." This field is a branch of mathematics that deals with the existence and character of solutions of dynamic models of various kinds. There are various branches of this broad field of inquiry; various terms from some of them are beginning to creep into common parlance. One widely repeated term is "catastrophe theory." Another is "bifurcation theory." Another one is "chaos," a mathematical phenomenon on which I will touch at greater length in a minute.

The possibilities for very complicated solutions to dynamic equations were fully appreciated early in this century by the great French mathematician and physicist, Henri Poincaré. He felt that the classical equations of celestial mechanics would never be completely understood because of the potential richness of solutions that was possible when there were three or more interacting bodies, even though the Newtonian system is completely deterministic.

In the '30's, Birkhoff pinned down some of Poincaré's insights. He discovered that the classical equations could exhibit many different possible periodic patterns, an infinite number. There wasn't just one stable solution upon which the mechanical system always converged. Instead, a given mechanical system might possess an infinite possibility of orbits and it might not converge at all, following a complicated, unpredictable, wandering pattern instead.

In the last ten to twenty years, there had been an explosion of progress in mathematical work of this kind, dealing with nonlinear systems. Properties that lead to these complicated phenomena are primarily the result of nonlinear interactions, which may be called "effect-reversals." In many structural equations relating one variable to another, reversal of effect is obtained. Let's say Variable A influences Variable B. As you increase Variable A the initial effect is to increase or stimulate Variable B, but after you pass some threshold level the effect is reversed: instead of stimulating, you depress Variable B. The causal effect is reversed. This property of nonlinear systems brings about much of the complexity for which we're getting a growing appreciation.

I began to generate results of this kind in deterministic stimulation models a long time ago, before the new mathematical understanding was known or being discussed. But because this new understanding has been forthcoming, it's now possible to go back over some of this early work and to develop an explanation of the simulation results that we generated. The explanation is easiest to understand <u>not</u> in the context of these rather complicated simulation models however, but rather by looking at some of the simpler standard economic models which have a dynamical component to them and asking, "Under what conditions do these conventional models generate complicated behavior?"

One example of such analysis involves the IS/LM framework. This analysis was the creation of John Hicks, who attempted by strategic simplification to reduce Keynes' general theory of unemployment and money to a two-dimensional graph. By adding a dynamic adjustment relation, Samuelson made Hicks' model into a dynamic one. It turns out that, embedded within Keynes' assumption of how the world works, are nonlinearities that lead to effect-reversal. This effect-reversal is in aggregate demand. It happens because of interactions on the money market--as demand rises, the demand for investment goods and consumption goods is stimulated because of rising incomes. But if the money supply is fixed, and incomes continue to rise, it causes a shortage on the money markets; then the interest rate required for the capital markets and the goods markets begins to rise drastically and that chokes off investment demand and precipitates unemployment. This is basically the Keynesian story in its simplest possible terms.

What I did was to look at some specific equations that would represent this phenomenon. It was then possible to apply the mathematical results to show that, in principle, there do exist situations of demand for consumption goods and investment goods and for money which would lead to unpredictable oscillations in the economy. My student, T. Y. Lin, then simulated the results on the computer to see if, in fact, the model would exhibit the phenomenon that had been proved to exist. His results showed that the computed model doesn't settle down to cycles of uniform periodicity--it wanders. What we are asking here is: with deterministic assumptions about the way supply and demand work, but with no random shocks, can we generate an empirical phenomenon, the phenomenon of erratic, unpredictable oscillations? And we can! What it suggests is that one can begin to gain some sort of understanding of why the world does these "crazy" things, why it does whatever it does in such an unpredictable way.

The question that arises now is whether this is a common feature of models of this kind. Is it regular? Would it happen with low probability or high probability? It turns out you can show that it can happen, and with high probability under certain conditions--a finding recently arrived at in collaborative research with my colleague, Wayne Shafer.

III. Toward a Theory of the Evolving Economy

I want to turn now from <u>technique</u> to <u>theory</u>. Last night, Sir James made a somewhat gentle and good-humored, but nonetheless a rather sharp attack on economists. As one of that discipline I began to whistle and think of all sorts of nasty remarks and muttered to myself several hours afterward, actually. But I got to thinking about it: why should I engage in rebuttal? The fact of the matter is that economists do richly deserve a lot of criticism--not about the nature of economic theory as such, but about the way that some members of our profession have applied perfectly good economic theory to empirical phenomena to which that theory is inappropriate. It is precisely in the area of dynamics and change that I think theory is most often inappropriately applied. In fact, the fashion of neoclassical economics has gone so far that some economists explain the business cycle itself as an equilibrium phenomenon. There is a really deep and important issue here. Are these things that we observe, these random oscillations, for example, to be thought of as equilibrium processes or as disequilibrium processes? How should we be thinking about economic change?

Of course, scientists always continue to use the theories they've got until they are convinced of different ones. Because we do not yet have convincing theories of disequilibrium, the tendency is to continue using old theories, even when they fail to explain. We all recognize that economic theory is badly in need of development. Should we pursue the equilibrium or the disequilibrium direction? The trend of my own thinking has been towards the disequilibrium direction. I do not have time here to develop my ideas at any length, but let me leave you with an abstract.

Our first step is to view the path of economic development as a sequence of states, each member of which has evolved from its predecessor in response to actions taken by the economic agents involved. The agents respond to their changing environment, taking into account their currently perceived constraints and objectives based on information inherited from socioeconomic and natural environments. Their actions, based only in part on rational plans, influence the system through a feedback structure that is only partially understood and is imperfectly accounted for in the planning process.

Because decision makers have imperfect information, limited foresight, and changing preferences, they cannot make plans that are consistent and optimal. In general the plans are imperfectly coordinated and are based on expectations that are seldom fulfilled. They rarely lead to equilibria in which supply equals demand and in which technical and social efficiency prevail. Instead, plans are inconsistent among the various agents and actions must be based in part on adjustment procedures or behavioral rules that have worked satisfactorily in the past and which may be revised in response to new information.

"Disequilibrium" institutions and marketing mechanisms make it possible for the individual agents to continue functioning when their plans cannot be fulfilled. Retail stores function as inventories, on display, that mediate the flow of supplied and demanded commodities without the intervention of centralized coordination or of complicated and time-consuming market bargaining and negotiational procedures. Banks and other financial intermediaries regulate the flow of credits and debts that facilitate intertemporal exchanges without simultaneous bartering of goods. Ordering mechanisms with accompanying backlogs and variable delivery delays, together with inventory fluctuations, provide a flow of information that facilitates adjustment to discrepancies between commodity supplies and demands. Insurance, and other transfer schemes such as unemployment compensation, place resources in the hands of agents who without them would have no viable course of action.

These agents and mechanisms of disequilibrium were not always so prominent. Barter and auctions were once used extensively and money consisted of "real" goods. If primitive economies had been stable and evolved toward equilibria, even crude disequilibrium mechanisms would have sufficed and then withered away, gradually becoming obsolete. Instead, those economies were unstable; new and more elaborate institutions have arisen.

In the real world discernable epochs, identified by characteristic economic and social activities and by characteristic resource inputs, unfold typically in a progression of growth, irregular oscillation, decline and more or less rapid and final demise. In the process crises emerge that threaten the working of the system. If it is to survive the system must generate new organizational forms. Thus a <u>creative intelligence must exist</u> whose function is the design and establishment of new rules and organizations that will avert the crises and set the system in motion again.

The creative intelligence which lies behind cultural evolution must be as old as man. Because of its function, humans alone among living things on earth transcend the blind and profligate forces of biological evolution. Genetic variation and selection is augmented by cultural adaptation and selection. Once set loose it continues to work whether in immediate need or not. It creates a cadre of seers and inventors, then scientists and engineers, who busy themselves creating new ideas, new organizational forms, new technologies for material transformation, all of which can be set in place when and if needed and prevailing conditions are propitious for institutional change.

As specialized production and exchange grow, new institutions or enterprises will be introduced to mediate the new disequilibria in supply and demand. These institutions and enterprises will buffer and ameliorate the imbalance that precipitated their adoption. At the same time, they introduce new decision-points--often requiring new decision makers (new agents), new information requirements, and new delays. This implies a new level of complexity and a potential increase in instability.

The new system will work well for a time precisely because of new technologies, new rules of behavior or new organizations. But it must eventually pass through the typical pattern of growth and oscillation. Through its growing complexity it has increased its inherent potential for chaos and crisis. It seems inevitable that this potential be realized.

We must suppose that people, caught in a matrix of shifting fortunes due at times to rapid or catastrophic shifts in economic opportunity and constraints, will not always stick with past organizations. They will search for new rules of organization and behavior, selecting some from the large inventory made available by creative intelligence. They will, as a result, introduce new mechanisms with which to govern adaptations.

Economic development may be likened to a fugue--blending in dialectical counterpoint the themes of novelty and growth, of demise and decay. Cultural evolution more and more becomes a frenetic race between the human faculty for creation and the disintegrative forces which intellectual artifacts eventually unleash. The outcome of the race, its goal or destination--if such exists--can scarcely be guessed. To run this mysterious race, it seems, is man's fate: La Condition Humaine.

IV. Implications

Note that I have not discussed "capitalism" or "socialism" in any of their theoretical or actual forms. This is because the theory I have outlined is not institution-specific. Rather it is intended to characterize any system of human economy. Nonetheless these are some concrete implications.

The first implication is that such systems as exist today will evolve. Some may disappear in the process, as is common in history and prehistory. New ones will very likely emerge. And this evolutionary process will not end in some final regime with fixed institutions and with fixed rules of decision and exchange. Communism will not bury capitalism; it will bury (and is burying) itself. Likewise, the form and content of capitalism will itself change, but its ultimate goal cannot be discerned because, according to this theory, it does not exist.

Indeed, democratic capitalism is best conceived as an evolutionary process in which the form and content of system components are allowed to change, both *in principle and in fact*, in which conflict is expected, and is explicitly provided for in conflict-resolution institutions of various kinds, and in multi-layered mechanisms by which social values are correlated with private interest and by which social goals themselves are shaped and modified in response to contending forces. Of the economic systems now in existence democratic capitalism most clearly accords with the principles of economic evolution. For this reason it may have a kind of superior, meta-viability, in which some versions of it will continue to appear--but only if it continues to allow for creative destruction, not only of goods and techniques of production, but also of decision-making procedures and organizational design.

From the long view, evolution, not "stationarity," is the fundamental fact of life. When individuals cannot adapt, they are killed; when organizations cannot adapt, they are disassembled. When societies cannot adapt, they are destroyed, swept aside, covered with sand, swallowed by jungle, or trampled under by a new regime. Humans are destroying virtually all mammal species with whom they lack a symbiotic tie. Industrialized societies are destroying the last vestiges of paleolithic man. The internal-combustion engine has driven away the horse, the oxygen process has replaced the Bessemer process. Everywhere and at every level, we see evidence that economic systems are unstable and that in any given form they eventually stop working. Just now, concern arises about the instability of the contemporary world economy. The emergent phenomenon that becomes most crucial for us is the next form of economic policy. From where does it come?

It must come from the same attribute of mind that guided the subtle hand of those stone-chippers of so many centuries ago who invented the spearhead; the attribute that makes it possible to create an image, to set in place a form in some way unlike any that came before. This attribute creates a menu of possible ways of life from which policies can be selected and behavior carried on. Once in existence itself, and perpetuated within the species, it always functions.

When something new is created, since structural information is changed, its full implications are seldom perceived. The hunting band may have led some innovative tribe to a flourishing culture, rich in food, ceremony, and leisure. Eventually, however, when the last woolly mammoth was butchered, the last herd of giant bison was driven over a cliff, a way of life came to an end. The inventors of finer spears, bows, and arrows unleashed a new spiral of progress--but one that eventually also succumbed in unexpected ways.

The implications are that: (1) intelligence creates new forms that at crucial times make possible a continuation of certain living groups; (2) with changes in structure, a people may persist for a while; (3) the ultimate consequences of adaptation are unknown, and the new system is also ultimately unstable, so that new forms must repeatedly be found. Morphogenetic man is thus an active participant in evolution, matter that has come not only to reflect on itself, but also to design the new forms it will occupy in its trajectory through space and time.

References

Day, Richard H. (1975). "Adaptive Processes and Economic Theory," in R. Day and T. Groves (eds.), Adaptive Economic Models, New York: Academic Press.

__________ (1983a). "Adaptive Economics," in R. W. Crosby (ed.), Cities and Regions as Nonlinear Decision Systems, American Association for the Advancement of Science, Selected Symposium 77, Boulder: Westview Press.

__________ (1983b). "Systems, Information, and Economics," in F. Machlup and U. Mansfield, The Study of Information: Interdisciplinary Messages, Princeton: Princeton University Press.

__________ (1984). "Disequilibrium Economic Dynamics," Journal of Economic Behavior and Organization, March 1984.

__________ and W. Shafer (1983). "Keynesian Chaos," Modelling Research Group Working Paper No. 8307, University of Southern California.

* An earlier version of this paper was presented at the 1980 Annual Meetings of the American Economic Association in Denver. A more recent version was presented at the Conference on Economic Systems Revisited - In Honor of Professor John E. Elliott, held at the University of Southern California, January 1984. In preparing this version I have included in Sections II-IV digested material from Day (1975, 1983a, 1983b, 1984). Space here does not permit a mention of the quite extensive related literature. The interested reader may refer to the foregoing articles for some of it, as well as to considerable more in the way of underlying theoretical analysis and supporting evidence.

5

MODELING THE SYNDROMES AND ACCEPTING THE DISCONTINUITIES

John D. Sterman

I think you will quickly be able to tell that I come more directly from the academic realm than from the policy realm but, as they say, it's outside the clash of opposing views.

I am going to try to introduce what perhaps will seem like a dissenting voice on the purpose of models in general, and global models in particular. That has to do with the proper goal of model-based analysis of foresight. I think there is a tendency, implicitly or explicitly, to believe that the purpose of models is the prediction of future events, discontinuities or otherwise--such as the dramatic rise of oil prices, or depression, or even what the GNP is going to be around election time next year. It's my basic thesis that such point prediction of specific events is neither possible nor necessary for effective policy making. Why isn't it possible? In my work on the dynamics of social and economic systems at MIT, we have discovered that these systems are subject to the influence of random events. When you take a model of any kind of system--a corporate system, an economic system, or a social system--you must include in it the continuous gentle rain of randomness: the fact that one day somebody gets sick, and on another day there is a strike somewhere, or somewhere else something happens that wasn't anticipated. These kinds of things happen all the time. But even that gentle rain of randomness, without major shocks, is sufficient to cause the specific behavior, the occurrence of events at specific times, which is unpredictable.

Somebody mentioned commodity markets. If you knew (or could know somehow) what the price of any commodity were going to be, of course you would take action which would cause that prediction not to come true. So in the case of social and economic systems, I believe that there are very fundamental reasons why it's impossible in principle to point out or predict specific events beyond a very short time horizon within which inertia simply carries the system forward. Over that short time horizon, because of the inertia of the system, you are not in a position to influence things in any event.

Now if it's not possible to predict specific events, then do we just drop our hands and go home and cower and hope that they don't wash over us too fast? I don't think so. I think we can still make effective policy, but what I mean by policy is perhaps not what everybody means by policy. For us, policy is a way of making decisions, and the purpose of policy analysis with models, formal models or otherwise, is to design policies--a way of running your organization or your system that will render you more or [sic] less subject, less vulnerable, to disruptions--and to make such disrupted discontinuities less likely. In that vein I see modeling not as an end in itself, but as a process. Nobody ever has taken a decision or been convinced of anything directly because of a model.

I am not surprised at all that there is no evidence of direct effects of global models on policy making. Instead, unfortunately, it seems that models are used typically to justify previously decided policy or position, to lend authority to an argument, or to provide a scapegoat when the forecast is falsified. Rather, we see modeling as a process, the goal of which is the improvement of people's mental models, people's intuitions, which are ultimately the basis for their decisions.

In terms of discontinuities, it seems to be commonplace among many people today that history no longer provides a guide to the future, because all the trends we counted on have been drastically altered. But in our view history is more relevant now than ever before and you have to go a little farther back than just the post-war era, which is typically the basis, formal or otherwise, for much of the analysis that goes on. History, in the minds of many people today, seems to be what was unemployment last year, what was GWP last quarter, or what Dow Jones did yesterday. We must take a longer-term view of looking ahead and looking back.

In our modeling work we have found that discontinuities should and can be modeled as the result of the interactions of ordinary ongoing policies of everyday people managing their everyday business. Discontinuous events arise out of continuous incremental change. You are never going to be able to predict with the model that OPEC will raise the price of oil by such-and-such an amount on such-and-such a day. However, our models, and other people's models, have been able to show that the context for dramatic price increases would arise at a certain time, that the policies everybody was pursuing with the best of intentions would lead to a crisis, and would lead to discontinuous change.

Let me give you an example. In our Limits to Growth model and the world dynamics project at MIT, we have been developing what we call systems dynamics, which looks at various limits to growth--in particular the energy transition question. But also we have been looking at what we call "the economic long wave" which many of you may have heard about under another name: the Kondratieff cycle, a cycle of economic expansion and depression over about a 50-year period. In fact, there have been major depressions in the industrialized nations: in the 1930's of course; in the 1870's and 1890's; and before, that, in the 1830's. Now, these are discontinuous events. They appear on many of the lists of discontinuous events which people have already presented. What we find is that our model explains how this arises out of absolutely mundane, even boring, statements of how ordinary people and corporations and governments are managing their everyday affairs. Let's take the example of the foreign debt process; I think we'd all agree that if there were massive defaults on Mexico's, Brazil's, and Argentina's debts, that would create a significant discontinuity. What our model shows is how such a situation could come about and it suggests conditions under which defaults become more and more likely and eventually less inevitable. We are never going to be able to tell you when and who is going to default first, but we would be able to say what conditions brought this about and what could be done about it at this point--although it might not be much.

So let me conclude. I hold with Tolstoy on this issue: that history, economic change, social change pertain to what individuals are doing on an ongoing basis. This is understandable with our current knowledge and tools. The purpose of model building is to improve our intuition so that we can endogenize the discontinuities, bring them inside the boundary of our models, not so we can predict them, but so we can take actions to minimize their effects and to make them less likely in the first place.

I want to expand on the idea that the purpose of modeling is not to predict the occurrence of specific events at a specific point in time, because that is not possible, and it is not important. I want to illustrate that by discussing the work that we have been doing on our systems dynamics National Model.

The National Model is the project that we have been working on for about ten years, since the Limits to Growth project. It is a national level model which we feel can deal appropriately with the kinds of issues and problems that are arising, all the developments in economies, and with the basic questions that we are asking, such as how the systems dynamics National Model can clarify current economic conditions. There are really a number of ways in which our National Model helps us understand what is going on.

First of all, there is the idea that there are multiple modes of behavior in the economy, there is more than one thing going on. For example, there are business cycles like the four-to-seven year cycle of expansion and recession which gets the overwhelming amount of attention in the present policy discussions. There is the long-term inflation mode and relative expansion of government, which has been continuous since 1789. There is the economic long wave, which I am going to talk about a lot more in a minute; and the life cycle of economic development, which is a long-term, 200-year process of take-off, rapid growth, and then, approach to either social or physical limits. These multiple modes are integrated into the National

Model; distinct mechanisms that create them are represented and we see how they interact and amplify one another.

The second phenomenon is what we call "syndrome recognition." Today there is a bewildering array of different economic symptoms--the economy ranges from high unemployment, to declining or zero productivity growth, to low capacity utilization, to declining return on investment; and on and on. And we tried to see these symptoms as parts of a syndrome, in the same sense as a medical syndrome.

Last, the purpose of the model is not to predict the future of the economy, or specific future events, discontinuous or otherwise, but to develop an understanding of where the syndromes come from and how they can be affected by policy.

The idea of multiple modes of behavior is that the complex social and economic systems in which we live necessitate multiple patterns of accommodation to complexity, and to understand what is going on it is crucial to understand where these different modes are coming from and how they are interacting. Today we feel that the three most important modes are the business cycle--a four-to-seven year cycle; the long wave--that approximately 50-year cycle; and what we call "debt motivization." The latter has to do with the development of a structural deficit, and the monetization of that deficit and consequent inflation.

I am going to focus on the long wave because it is the mode that we feel is dominating developments in the economy today, and the one which is least understood. This Kondratieff cycle (a 45-to-60 year cycle) is characterized by about thirty years of steady expansion and a climb in the rate of growth, a peak phase, a downturn, a transition (which is our euphemistic way of saying "impending depression"), and then the trough, or depression phase. Historically, there is some uncertainty about this; scholars who have looked at this agree that the depression periods, the downtrend periods were the 1830's and early '40's; the 1873 period to the late 1890's period; the 1930's, of course; and today.

Let me discuss some specific data. I can't spend a lot of time on data here, so I don't expect that this will be a convincing case, partly because the data are a little equivocal. There are data on real GNP in the U.S.--in constant 1972 dollars, from 1800 to the present, and, of course, the dominant trend is the 3.5% exponential growth in GNP and massive growth in the economy--so much so that you really can't even tell what happened before 1900, because it is on such a small part of the scale [sic]. The data do show the Great Depression, where growth was above the trend line from 1900 or so up to '29, and then was below it. In the '30's it pops up during the war and then it's sort of back on the trend line and then above the trend line during the '50's and '60's, the "good ol' days." After that the data show a growing severity in the business cycle, followed by a significant departure from that trend in the last three or four years. So there is something going on here. Something else happened, too: there was a serious set of depressions between 1878 and 1890.

Now this is not intended to be a historical validation test, but just to note that our National Model is generating the different kinds of behavior that you see in real life. In our model, we have the long-run growth, the same as we note in history. Second, we have the long wave, we have a depression, around the time of what was called the Great Depression. This depression takes place around 1880 [sic] in the model, and a graph will have little bumps which are clearly visible: the business cycle, or short-term cycle. Even though the business cycle gets the most attention, it is really the least important of these different modes. Now, it is hard to say that there is a long wave going on there, but other data will show it a little more clearly. Consider the consumer price index (historical data from 1800 to the present, approximately), which shows a fifty-or-so-year pattern where consumer prices fell between or after the war [sic] to the start of the Civil War. They jumped up and they fell; then they jumped up and they fell again; once more they jumped up and there was a shift--like a discontinuity you might say--where there has been massive inflation in the last thirty years.

What we have, then, are at least three different, distinct modes of behavior. First of all, we have seen long-term inflation--something new in the last thirty years; second, we have the long wave; and third, we have the little bumps which are price changes caused by particular events during the short-term or business cycle.

Well, those are the data, and lots of discontinuity is at work here. In our model we have the same three modes: long-term inflation starting in the last thirty years, the long wave in prices, and the short-term squiggles caused by the business cycle. Now there is no exogenous variable in this model except for noise --what I call the "gentle rain of randomness"--so that everything else, especially the long wave, is being internally generated by the National Model. Another piece of historical data is the unemployment rate from 1890 (which is about as far back as you can get reliable figures), up to last year. Unemployment went up to 25% in the Great Depression. Before that the next largest surge was in the depression period of the 1890's, when it went up to 18%. Since the Depression, of course, the highest unemployment rate that we have had is the present one.

Over a longer time span than the other data represent, our model shows short-term fluctuations caused by the business cycle and by small random shocks, and the same major surges in unemployment approximately fifty years apart. Now, they don't all look alike; every long wave is different. But you've got a systematic recurrence of high unemployment, approximately fifty years apart. When I said that discontinuities ought to be brought into the boundary of the model and ought to be made endogenous, I wasn't suggesting something that's impossible to do. Ours is a model which generates a wide range of discontinuities internally, endogenously, and from a micro foundation. The kinds of equations, the kinds of representations of human behavior in this model are very ordinary, very mundane; they describe ordinary management and everyday affairs on the part of firms, individuals and governments. It is not an econometric model--there is no sort of macroeconomic type of equations such as are present in typical econometric models--it is a very micro foundation approach.

Interestingly enough, in the runs which I've discussed, there is no international activity whatsoever. You may say, therefore, that this model is crazy, but our point of view is that, clearly, we live in an international environment, where international trade and debt and so forth are an important part of the current crisis. Nevertheless, the model generates a very wide range of the symptoms that the economy is experiencing today, which suggests that, to a much larger degree than is typically perceived, the problems we are experiencing are internally generated, internal to our economy. Now, that is a good news/bad news situation. Bad news is: the problems are our fault; good news is: we then have some leverage over them, we have the ability to influence them.

Let me talk a little about this lack of predictability. I have argued that, in general, cyclical behavior modes (other behavior modes also) are not exactly periodic. Economic systems are not like pendulum clocks that repeat exactly. Instead, they're highly influenced by short-term random shocks. What this means is that there's a limit to the accuracy of forecast--I think it's a very fundamental one--and it means that the burden on policy makers is to understand what's going on and not to predict. Let me illustrate that.

The model generates the long wave. It explains what's going on today across a very wide range of phenomena, but we can't predict exactly when the next drop in GNP will occur or how high unemployment is going to go. That's perfectly consistent with history. There seem to be at least two kinds of long range transitions. First, historically, there's been long stagnation with frequent severe recessions and weak recoveries. The best example of this is the period from 1873 to 1897. On the other hand, there also seem to be long wave downturns that show up as sharp sudden catastrophes, followed by depression. Industrial output per capita from 1920 to 1940 just followed population growth, and there were little business cycles from 1920 to 1929; then there was a precipitous collapse--down to '33--and a long, slow recovery, with another recession in 1937. By 1940 we were barely recovered to the 1929 level. That's the cataclysmic end of the scale.

The other end of the scale is industrial output per capita from 1860 to 1900. From 1837 to 1897 was a global period of depression but in the U.S. there was economic growth per capita over that whole period. However, we had one spate of major depressions. Indeed, the one in 1873 and the one in 1893 were called "great depressions" because they were so much more severe than what came before or after (until 1929).

So that's the range of expanse that you can have in real life. The model generates the same kind of variability, simply due to the continuous rain of randomness. That means that we can't predict exactly what's going to happen. We can predict "the continuing transition or continuing stagnation" over the next approximately ten years. We can identify particular areas of stress and particular danger zones of international debt (protectionism being the most notable) which are likely to intensify and stay with us.

Let me now suggest a couple of ways in which our modeling approach differs from ordinary economic models. First of all, we do not assume rationality. Several speakers commented that the world isn't rational--people don't make decisions in a rational way. I took that to mean that people don't make decisions according to the optimizing profit of [or?] utility, maximizing models with good information. That's usually assumed in neoclassical economics. We don't assume that, either. Instead, we follow in the work of Herbert Simon, which is the theory of bounded rationality. Our basic viewpoint is that there are limitations to the information-processing and decision making abilities of real people. We try to take those limitations into account. That means that people can't take a God's view of the whole system. It's just too complicated; instead, they break down the problems of managing their firms, managing their personal lives or governments into separate subproblems, each of which becomes more simple because you ignore what's outside that little boundary. The result of that is that people have simple rules-of-thumb which they use to make decisions and that they become what is called "intendedly rational." They are trying to do their best. They have very imperfect information--information that is out of date, biased, imperfectly measured, inadequate--and consequently they don't have the ability to do complicated optimization to what's going on. So we model people's decisions in terms of these kinds of relatively simple decision roles which are supposedly rational; people are doing things that make sense at the micro level. But when you put it all together--when you allow these individual, intendedly rational decisions to interact in the context of the whole system--you can't get globally rational behavior.

Now to get specific. In the long wave we feel that behavior is basically generated by inherently ancillary structures in the economy which come about from the ordinary management of firms [and] which are magnified by a variety of reinforcing processes revolving around capital investment, labor markets, financial markets, interest rates, and consumption/savings decisions.

On the investment side, we deal with what we call "self-ordering," which is that if there is an increase in demand for goods in the economy, then the producers of capital equipment receive a higher volume of orders from other industries (or they try to expand) and those producers of capital equipment themselves then have to expand. So they need to order more capital, and that further expands the demand. Now this is nothing more than an accelerator mechanism to the economist, but the difference between our model and the economic model is that we look at the disequilibrium pressures that result.

One of these pressures is an increase in the demand for capital goods which runs the inventories of these producers down and increases their backlogs. So they have to produce not only at the higher current rate of demand, but an an even higher rate, in order to bring those backlogs down, to build their inventories back up. That means they need even more capital to increase their production capacity so they place orders for more capital and that further expands the demand, further depletes their inventories, and further boosts their backlog. So we get disequilibrium pressures which reinforce the original expansion in demand even more.

There's a wide variety of these swings. The need for capital goes up and you can't immediately satisfy that demand because it takes time to acquire new production capacity. So the lead time, the delivery delays time is lengthened; that means people must plan further ahead. They have to order further ahead, and that further increases the demand.

We may be dealing very much with an economic model but it's not using neoclassical economics. In agriculture, the hog cycle is very much a part of this. The self-ordering idea in hogs is that it takes hogs to make hogs, and that's one reason why the hog cycle is as strong as it is. That same self-reinforcing mechanism destabilizes the market in the same way. But the [economic] lifetime of hogs--the hog cycle--is about 48 months. The lifetime of hogs, the [economic] gestation period is a couple of years. The lifetime of capital equipment is twenty to fifty years and the comparable cycle that is generated is fifty years.

Another self-reinforcing process that helps generate the long wave has to do with the build-up of a growth culture. Growth expectation soared during the '50's and '60's and growth became the thing to expect. When there was a temporary downturn, people didn't let that affect their belief that the economy was going to continue to grow at that 3½-to-4% rate. Expectations of high growth lead to a greater desire for production capacity, so that leads to more desired capital, to more orders, more investments, and finally becomes a self-fulfilling prophecy. In fact, that demand materializes, builds up and gives momentum to the expansion of capital.

When I say that it's not a neoclassical model, I'm suggesting that we are not presuming equilibrium or optimizing behavior. We are presuming that people are trying to do their best and that, given time and without any random shocks, they will, in fact, reach the neoclassical equilibrium. The difference is that it takes time to discover what that is--there is need for a learning process model. We don't assume that people can sit back, look down on the whole system, and say, "Aha, now I know what to do." They have to discover answers through a process of trial-and-error and through experimentation, and that takes time. So there's a very large difference between our stance and the neoclassical approach. We are not assuming this optimizing ability; we are not assuming perfect information. Instead, there are realistic assumptions about how people go about making those decisions. At the same time, we must include the price and market mechanism and all the substitutions processes which are present in a neoclassical framework. It's just that we are not assuming that you have already come to a perfect state.

6

REALITY, INTERACTION, AND DISCONTINUITY IN ACCEPTANCE OF MODELING APPROACHES

Lionel S. Johns

I have been making notes as I was listening to the earlier speakers, and the first thing that I have written down is that we treat models as an end in themselves, rather than a means. I recognize that John Sterman said something I thought deserved saying twice. In other words, I feel that I have to defend the policy makers who aren't here tonight with regard to the constant plea of modelers that policy makers should pay more attention to the modeler.

The policy makers' answer is very straightforward: they are very busy making policy--whether they have knowledge about policy they are making or not. The urgency of making policy is so great that they don't have time to learn about making models. As a result, if you are going to bring to policy makers a product which emanates from a model, don't tell them, because in Washington they hate models, and the way your model would be characterized would be expensive, mysterious, irresponsible, never right, always innocent, and always manipulated by the modeler. When there is a problem with what the model says, the modeler insists it's the model's fault. The result has been (and I'm really quite serious) that the policy makers do not want to hear about the output of a model and they also don't want to be told that one should learn and understand models and their assumptions so that the government can use them as tools.

First of all, let me offer what seems like an appropriate anecdote. My office works for a Board of Directors: six senators and six congressmen, equally divided by party and by a wide spectrum of political perspectives--from Sen. Orrin Hatch on the right to Sen. Ted Kennedy on the left in the Senate, and in the House, Rep. Clarence Miller to the right and Rep. Mo Udall to the left. That wide political spectrum is very useful for us. But what I am indicating is that, despite diversity, we proposed the largest project we had ever attempted. We called it "Technology and Economic Transition." It is aimed at understanding, in a large-scale or strategic way, what the direction of the economy may be for the next five, ten, fifteen years and what the implications are for public policy. We are constantly asked whether certain transition or change going on in the country should be stopped, facilitated, or whatever. It seems that what is most likely to happen is that policy will be made at the wrong time and problems will be aggravated, instead of solved, by such things as local content legislation. In describing the aforementioned project we made the mistake of saying to the Board that we would construct scenarios, that we would look at a synthetic future, that we would compare it to an econometric model output. Somehow they took this to mean that _we_ were going to _build_ a model and the project was very nearly disapproved, until we could finally point to the sentence inside the description of the project that said its purpose was _not_ to construct a model. As far as the Board was concerned, if we were going to construct a model, it would never be completed, it would cost much more that it was expected to cost, and it would never be useful to anybody. You really have to use the term _model_ very carefully in Washington.

But I'd like to come back, if I may, to the description that was given earlier about what brought our host Group together--not so much to look at _Global 2000_, but to look at problems of a global nature and the detection or anticipation of problems: how to understand their character, and how to avoid or mitigate problems that might arise at the global level by anticipating them. The answer is not to build a model--the answer is to bring about some measure of understanding. If you use a model or models as instruments for that purpose that's fine--but what you are aiming at is achieving understanding, or insight, or foresight, by the policy makers, so that they may be aware of what is likely to happen in some comprehensible way and take appropriate action should it be possible. That seems to me to be the aim and theme of this conference.

I would like to offer an analogy--something that took place last week. Paul Ehrlich and Carl Sagan came to Washington with a group of responsible analytical people, principally from the West Coast. They described a result of nuclear war which had not been anticipated; in fact they produced a report on the effects of nuclear war delineating a major impact never identified before, which this group had been analyzing for more than two years (and had subjected to substantial peer review). They came to Washington and they held a conference; they made no policy comment. The conference was intended to present the technical facts as best these scientists could understand their analysis. Of course they had used global models because they were looking at circulation, climatology, and ecological effects. What they reported I found incredible, although I am not here to sell the thesis. They start with the premise of an exchange in nuclear weapons at some threshold above 100 megatons (well below a likely exchange). Such an exchange would introduce a loss of light in the northern hemisphere as great as 99.8% and no less than 80% for a period up to or in excess of a year--one which would create a temperature loss on the order of 30° to 50° C after a few days and would remain for that entire period. The ecological effects resulting, after prompt effects would kill as many as 160 million people in the United States, suggest that you may lose most of mankind in the northern hemisphere.

The new knowledge which these scientists revealed is that this damage would be caused by smoke: the darkness would be caused by smoke, the temperature drop would occur because of the opacity of the smoke in the air. Consequently, new radiation would be distributed and would be circulated in both hemispheres.

Those findings of what a nuclear exchange would bring about offer a view very different from what anybody in Washington had previously realized. The analysts had before them the products of models; they came to Washington, they gave their talk, they had a live hookup with Russia, they had T.V. cameras present from Britain and from France (where the entire conference would be shown in its entirety) and, they did not permit political discussion on the podium about the implications of their findings.

What happened following that report? Let's talk about some of the forthcoming policies that might be implicit if the data were accepted. First of all, a primary warfare fighting posture would no longer be valid if these findings were true. I mean, the idea that you would have a nuclear exchange, that you would have a strategy for an exchange, or even that you would have a counterforce attack on your enemy's strategic weapons, is now out the window--because if what they say were the case, a counterforce exchange of approximately 1000 megatons would kill the "victors" eight days after they destroyed the "losers." You would have destroyed a portion of the counterforce of the first strike capability of your enemy and the damage that you would have done to the atmosphere would have destroyed much of your own population. For example, if the United States made the first strike, and if the Russians were to counter, then surely that threshold would be crossed and everybody would lose--and lose in a much greater way than had been anticipated.

What has happened since this talk was held in Washington and since the product of these models has been delivered? There was little testimony on the Hill. An account was published on the front page of the Washington Post, some of it was in Parade, but there have been few hearings. Whether DOD believes the presentation and will be investigating it--and considering the design of alternative strategies for nuclear warfare that may be a little less suicidal--is an interesting question. Are they contemplating whether you should buy 21 more MX's, or 200 more, or 2000 more, if we already have, say, 12,000 megatons (way beyond threshold) in the collective strategic arsenals? Is it rational to buy more if you believe this strategy?

What I am trying to show is that there is a rich content of policy implication here in the event that the report of these scientists is true. Last April [1983] they subjected the report to several days of peer review at MIT, and the predictions stood up to this review in quite a substantial way. Consequently the findings of Sagan and associates aren't just a flaky idea.

The point in my giving you this tale of woe is that if you believe that one can sit some distance from global policy making and develop a model and get the policy makers to pay attention to it, you have not put your arms around the whole problem. If your intent is to get your findings into the forms in which they will be heard and debated, and that policy will be made with those insights considered, then there is a lot more to do than simply develop the model. How do you go about having impact on the process? I suggest that the way to have the impact that you are seeking would be to build backwards: inquire into the kinds of insights needed, the kinds of models needed to provide them. In that way you have a "market" for your output in the policy arena and can tailor your product to meet that need.

In the academic environment you're accustomed to professors who stand before students and share their wisdom. But in Washington one is accustomed to the notion that there is much more wisdom in the audience than on the podium. For that reason the format that we tend to use is one in which there is more audience participation than we've seen here.

I have found, in listening to the discussions, that I'm in the discontinuity business and didn't realize it: I just didn't know what to call it! The oil shock, the oil disruption is one case in point. We knew that there was some prospect of a discontinuity in the region of the Straits of Hormuz, and that this discontinuity could manifest itself in six to eight million barrels of oil a day not being able to make their way to the market place. We knew that we didn't want to get into purely economic questions of how to allocate strategic reserves, or how critical it was to build them up at a faster rate. In effect, we didn't want to forecast optimum means of distribution or argue whether the market could make such distribution--or whether you should say, "Well, the hospitals can't bid very well for oil, so you should deliver oil to the hospitals and not make them bid for oil in the market place." We wanted to set those questions aside and still envision that all of those problems would be resolved at some juncture.

But there is a larger question implicit here: if there is a disruption, the vulnerability of those oil fields to physical damage could cause production to go off-line for quite some time. While $100/barrel oil could, in the short term, mitigate demand--at the same time it wouldn't necessarily preserve a robust economy. The larger questions would include: what policy changes might be considered for preserving a healthy economy? What sort of federal policy could enable us to reduce the impact of an oil shortfall for some considerable period? This leads to such points as finding out that electrical utilities, sitting side by side, are not connected to each other: some of them have a lot of excess capacity and some of them would need additional capacity--some are highly dependent on oil, others not. In fact, you could hook two together and both might benefit. You might never use the hook-up--but in the event of disruption one region of the country could have the impacts substantially moderated by using excess capacity from another area. There are ways to facilitate the providing of energy from conventional domestic forms through segments of industrial society that would otherwise not have access without the removal of some institutional barrier. One must face actually analyzing potential discontinuities and looking for ways to mitigate those discontinuities. That's something nations must do on a regular basis!

Now what I would like to argue (I'm amazed that I'm making this argument) is that extrapolation is all right insofar as discovering discontinuities is concerned. It's the ability to deal with the answers <u>within</u> a model which is the real problem. If one takes discontinuous change and asks, "Why should one care?", we acknowledge that from the policy point of view, the goal is to avoid political instability--whether this be as minor as a politician keeping his job or as major as a democracy maintaining its form of government. A second goal is to reduce or avoid human misery. There's a third goal--one which we frequently overlook--and it is the detection of those discontinuous possibilities which are good news. Innovation may occur as break-outs, as break-throughs, or as significant course changes. They may indeed solve societal problems, either existing or anticipated.

We would all like to recognize that discontinuous change can be beneficial but the momentum of society, and its institutions, may keep those things from happening before they ever have a chance to benefit that society. What one would like to have happen, as these discontinuous changes are recognized, is the opportunity to change course, so that one can either avoid potential chaos or reap the benefits of new opportunities. One seeks to mitigate the effects of the discontinuity by reducing human misery or preserving political stability.

Now, where do you find the seeds of discontinuous change? I think we get preoccupied with what you might characterize as "out of the blue." We are obsessed with the notion that we must continuously search the horizon because some event is going to spring on us and we don't know what it is. We assume any such event will be bad and we won't have good sensory mechanisms to detect it. I would argue that there are almost always large segments of society which perceive discontinuous change is underway before society-at-large reacts to it.

One of my favorite examples is General Motors. Detroit was the last sector to hear that car preferences were changing in the United States. No attention was paid until maybe five years after such changes had taken place, but any man on the street could have told you what was going on. In the same vein, if you take the nuclear description I presented earlier, there is a segment of our society convinced that a substantial discontinuous change is lurking on the horizon in the form of nuclear war--yet the political leadership in our society will not alter its policy. Those who are policy-makers will not alter policies until our democratic process works its will; the public must force debate of those issues by its elected representatives.

If one were to describe what it is that brings about discontinuous change, I would argue that it is generally technology-driven, policy-driven, or value-driven. For example, we spend a lot of time on natural events rather than on the technology-driven or policy-driven aspects which continually change. For example, if you were to look at a trade policy which was in retaliation against Japan for trade imbalances which exist today, you could make a strong economic case (in the short term) that the disequilibrium between exports to the United States and imports from the United States is unacceptable. However, if you were to set a group of policies in place that quickly equilibrate this imbalance, you would put Japan in a very serious state of economic depression. I would assume that the Japanese are capable of considering this as a possible outcome. So one would have to ask whether that is a potential discontinuous change--something that will affect us very unfavorably, instead of favorably, as the policy would intend.

Suppose we consider what Japan's strategy might be in this hypothetical case. Look at the Pacific Basin today: most of the nations in the Pacific Basin don't trade heavily with the United States--they trade with Japan. The raw materials flow from Australia, Malaysia, Thailand, and Indonesia, and most of the manufacture that takes place in Hong Kong and Thailand and elsewhere is all part of a fabric which is the Japanese trade network. Obviously, our hold over the Pacific Basin nations is actually the protection of our military umbrella.

Another policy the U.S. is pushing very hard (and for good reason) is that Japan should invest more of its economy in defense, and in the defense of those countries with which Japan is concerned. That this would destroy the principal strength of the United States' ties in the Pacific Basin seems little considered.

Suppose we pursued the two policies I have mentioned: first, that we deeply cut Japanese imports into the U.S., and second, that we succeed in getting Japan to increase its defense spending? Japan could not permit the United States to shut down its economy. If the yen were relatively well protected from world monetary markets (which many will argue is now the case), and if something like a Marshall Plan were created by Japan, in which substantial credit would be extended to the Pacific Basin nations, the flow of the economy would then be diverted to those nations, including western South American nations and certain Central American nations as well. If that were to happen, Japan would have markets to keep its economy functioning--it would create a yen-based economy and its extension of credit could in fact create a whole new economic and political order in the Pacific Basin.

If the two policies I have noted were to be pursued and such responses were forthcoming, would that create a discontinuous change? The answer is, "certainly!" and it might precipitate some human misery as well. We seldom explore policy for whether it may create discontinuous change. The Japanese bombed Pearl Harbor, following a U.S. steel boycott. To what degree did that policy contribute to Pearl Harbor?

I believe we need to look much harder at the degree to which discontinuous change evolves from a policy, or a series of policies. We must begin by asking how a modeler goes about assuring that he can understand, detect, or define continuous change. One has to rely on the people who are most involved in whatever it is that is being modeled. For example, if you ask any of us here to identify which developments in biotechnology are going to alter the way we live--whether it will be from a new source of food or resistance to disease or from extended longevity because of genetic manipulation--we could only make wild guesses. But those working in the biotechnological community have a pretty good notion of where the breakouts are likely to occur, and that means both hazardous and other significant events. If its technology is being modeled, the biotechnology community involved must be consulted in order to recognize the possibility of significant changes. The modeler then would not be surprised by new events because he would have been informed in advance.

The real problem, I think, is that the rational basis on which you construct a model is one which presupposes that the answers are always reasonable. We keep the model from flying beyond the bounds of reason by continuously functioning in such a way that the outcome is one that is not unexpected. If that is the modeler's intent, then those explicit formulae or factors are going to be put in place which will cause the model to keep from recognizing discontinuous change because such occurrences would look like "fly-up" in the model itself. My point is that modelers need not be preoccupied with looking for unexpected change; if they want to detect the probable areas of discontinuity they may call upon those who are closest to whatever area is being examined. That community of experts will probably predict very well where the latent discontinuity exists. The difficulty in modeling lies in looking inside the model and being willing to accommodate the results of discontinuous change.

7

PERCEPTIONS OF MODELS, GOLDEN KINGDOMS, AND POLICY FUTURES

Lindsey Grant

When I got this invitation, a rather irreverent little tune came to mind: "Though their skirts are made of gingham, it's the saucy way they swing 'em brings the drummers to my hometown..." Well, it was the saucy way you swung that discontinuity idea that attracted this drummer, because I think you do have an essential issue here. I may say some rather unkind things, because I have been asked to talk about the relationship of modeling to policy.

Let me preface my remarks by saying that I was on the National Security Council staff and later in policy planning in the Department of State. I spent a good deal of time being frustrated when we were supposed to be planning policy; the primary impression I came away with was the extraordinarily narrow tunnel vision through which we make policies. It was pretty shocking and I got more and more interested in the problem of how you widen your perspective. I think this goes far beyond modeling. It seems to me that the answer lies, as much as anything, in organizing the system--the techniques for bringing relevant knowledge to bear before we make national decisions. It's also a matter of getting people out of the crisis atmosphere and helping them see what may be coming a little further down the road. Modeling plays a role in this, but only a partial one.

When I hear modelers talk, I see a world like Tennyson's: "Let the great world spin forever down the ringing grooves of change...". Note the conviction that there are some <u>grooves</u> here. I think that's very Victorian! I am not sure we are, even now, convinced that there <u>are</u> grooves; we certainly don't know where they go. We might even turn it around and say that, for anybody involved in active policy decisions, the nature of the world basically is one damned discontinuity after another. If you people can find a few continuities and show that they really happen, you are way ahead of us. Don't worry if a few discontinuities trip you up; we are way behind that point.

In reviewing a number of big and dramatic events in which I was involved at one time or another, I asked myself: is there any way that systematic--presumably mathematical and computerized--models could have helped us in those decisions? Actually, I found very few cases in which I could even envisage a way they could have helped. I like to contemplate one very small but dramatic event, because I came out right. In 1974, the conflict between the Greeks on Cyprus and the Greeks in Greece, which led to the intervention of Turkey on Cyprus, was the product of an insulting and ill-calculated letter from the Cypriot Archbishop to Ioannides, the leader of the Junta in Athens. A British diplomat said to me that Ioannides could not do anything to retaliate; it simply would be irrational, because if he did set something in motion, he would have the Turks on his back. I argued that the diplomat was assuming rationality. Now Ioannides's pride, his "pholotimo," was hurt. He tried to assassinate the Archbishop, and various things happened that are now history.

At that juncture there could have been a Greek-Turkish war which could have changed history, and undermined the whole nature of the NATO alliance in the eastern Mediterranean. This didn't happen because another man (Karamanlis) followed the ouster of Ioannides in Athens. He was the one man in Greece with the authority to overrule his cabinet and to tell that cabinet that he would inform the press and the world that he did not intend to fight Turkey over Cyprus, and that he would say it was a joint decision by the whole cabinet. It was sheer good fortune that Karamanlis (who had been in London) was not Khomeini (who had been in Paris). He had a rational understanding of what would be best for Greece, and he made a decision that kept both Turkey and Greece on the side of the West--rational in the broadest sense, but perhaps personally adventurous because he risked his position by taking an extremely unpopular position at the outset of his incumbency. I cannot imagine a computer program that would have predicted Ioannides' fall, Karamanlis' succession, or Karamanlis' political power and statesmanship.

As for Ioannides, what was rational and what was irrational? Maybe he said, "If I ever let this fellow to get away with insulting me, I am going to be dished, because everybody will think I am a paper tiger." So perhaps he thought (rationally) that to survive he had to do the thing he did, which was totally irrational in terms of Greek interest. How do you make a model that will show when people will be rational or irrational? And at what level of rationality?

In other words, the problems of defining the nature of decision making are so diverse that I despair of creating a model that is better than a single human mind in generating a long-term description of the way the world will run.

I think one is in danger of determinism if one expects a computer to <u>predict</u>. If we really knew about a problem, we'd do something about it, and this would create changes which probably would not be anticipated in the computer model.

I might offer one other example to illustrate the role of the modeler as it might relate to the decision maker. I spent the last couple of days before I came here looking at several models now being run. One of them is an effort to model U.S. agriculture in the future. This model was formulated (I won't say by whom because I am going to be pretty hard) in a reputable university, by a reputable group of professors, and it appalled me. It ran three different projections for the future of U.S. agricultural production. Two of them concluded that we would be net food importers by about the turn of the century, and [or?] shortly thereafter, under even the most optimistic scenario.

This exercise committed the cardinal faults in the relationship between a modeler and a decision maker. For one thing, the modelers wanted a good, easily handled, easily understood conceptual premise on which to work. They decided to use energy inputs as their surrogate for the possibility of increasing agricultural output. They then decided to allocate to agriculture, throughout the whole period of the study, only the same proportion of total U.S. energy availability which it used in 1980. They made another assumption: that available energy would be assigned first to off-farm uses and then to on-farm uses--on the assumption that a crop wouldn't be raised if it could not be processed. The model had no price mechanism. It used three different assumptions about human fertility, and for immigration it used a dated census projection of 400,000 a year. (The Census can't count illegal immigration, so they simply ignored it.) Therefore what you have is bound to be irrelevant to policy choices. No matter where these people come out, the problem is that nobody is going to pay attention to them once they understand what the assumptions are. Even if the modelers say, "We are just running this--we just want to test it--we want to see what happens under these conditions," they have lost all credibility.

What I will be arguing is to turn this whole thing around. First, make it relevant to the policy decision maker. This model has no variables that relate to any decision that the policy maker would be making. In the real world, the coefficient for the energy input:food output ratio will shift. If we run short of food, or if world demand is high, the relative prices of food and energy will change, and so will the ability of agriculture to compete with other sectors for energy. The way that the model allocates energy within agriculture is, realistically, very dubious. Since the model runs mechanically, there is no way for the policy maker to ask it, "What if we change this...?"

Even on population change, the modelers elected to hold immigration to an unrealistically low figure, while they varied fertility all over the place--from 1.7 TFR to 2.7. The U.S. Government isn't ready to try to address American fertility. But it has been really wrapped up in the Simpson-Mazzoli immigration bill and the whole question of what to do about immigration. One thing that would have made this model relevant to current issues would have been to turn it around and make it show the implications of different rates of population growth resulting from different levels of immigration.

Another approach would have been to limit the assumptions and simply say, "If you want agriculture to grow so fast over the next twenty, thirty, forty or so years, here are the implications for acreage, for energy use, the rate at which you will have to use fertilizers, the rate at which you may have to increase pesticides, the requirements that you impose on genetic improvement." Then, when they are discussing energy, introduce a model to show how much agriculture will need. If there is concern about the increasing use of pesticides, show how a lower rate of application might affect other inputs and food output.

I now perceive a couple of things that I might not have been aware of prior to this conference. The first of these is the very real humility that a number of modelers have expressed about the ability of models to provide hard information for policy choices. This is not always what one hears from modelers. I agree with you. I think you affirm what I have sensed as an outsider.

I do detect that in periods of turmoil people invent "golden kingdoms" somewhere in the past. It's terribly important to us to have stability. Dr. Sterman has made this explicit. He's listed this as the first thing the modeler is trying to establish: stability. I would simply say, before I talk about the relationship to policy, that in a sense discontinuities are just frustrations because the world isn't the way you have in mind. You're trying to create that "golden kingdom" still, and reality isn't stable.

The next point I want to make is really a request or a suggestion. Before we leave this conference I suggest we write down our own definitions of discontinuities. We've thought about them a great deal; we've only made one attempt to define them. I find that attempt has been too broad because it is rooted in the perception that somehow there is a golden equilibrium and we're only momentarily distracted [<u>sic</u>] from it. We all know the degree to which this false equilibrium model is a feature of neoclassical economics and Keynesianism.

As I have said, the question of getting government to think better is not just a question of using these models. It is a lot of other things--only one part is to what degree you can use mathematical models. The politician, as much as you, is pursuing that same golden stability. And the first part of that stability is that the politician seeks to stay in office. I worked for Henry Kissinger, and one thing that we recognized about him was that he had absolutely no sense of the passage of time. We were right back with the Congress of Vienna: a brilliant mind reworking the same material.

The important issue is not to bring into the process the simple objective of playing mathematical games, but rather to warn people when there are intersections of trends--when the real situation is not stable, but dynamic. For this warning I must credit Barry Hughes' work. As far as I can see, discontinuity to an input-output modeler is nothing but standard sausage for the interactive modeler's mill. I think that we should go back and try, before we finish up this conference, to identify what we really mean by discontinuity.

Following the definition of discontinuities, we should identify the critical issues for the modeler in his relationship to policy. How can we best get the policy maker to look far enough ahead to see where one of these intersections, which we've been calling discontinuities, may occur? And for that, my argument is: don't spend the few minutes of conference with a policy maker in trying to make him see all the assumptions, the formulae and the coefficients which you believe provide a rational basis for projection. The policy maker thinks he already has a rational picture of the world. My advice is to take current trends and simply warn him, "If you go in this direction, you're going to have this kind of problem. If you keep on relying on commercial agriculture, you're going to be out here with the anticipated use of energy, you're going to be out here on eutrophication problems; this sort of problem will confront you on acid rain; this sort of problem for carbon dioxide. Do you want to be in that position? If you don't, here are options as to how you might be able to influence things."

Your advice to the policy maker might start from the present even though you know the present will not last. I would advise the modeler to keep current and exhibit the first range of implications. Only by doing this, only by using the extremely simple approach and never allowing the hearer to fall behind your reasoning can you really exert influence.

The final thing I would like to say is that I'm not sure that you really want to bother. Frankly, I am not sure it is terribly important to the people doing the kinds of intellectual modeling described in these sessions to be able to go directly to the policy maker. I think it is much more important that you do the kinds of things that Barry Hughes and Gerald Stairs mentioned--affecting the educational process. Through this process you may create awareness among sophisticated people who influence opinion in this country regarding certain looming issues. It may be just as satisfactory if intermediaries pass the message to the President, saying, "There are a lot of people out there who think you are going to have trouble if you do this." If the people "out there" can formulate the choices and their significance so that the people who write options papers for Presidents can understand, the effect may be considerable.

I would conclude by saying that I'm not so sure that the topic you brought me here to discuss is really so important. You may not need to reach the policy makers directly, but rather to build the informed consensus which eventually reaches them. The educational method may be the most significant way to prepare our leaders to meet discontinuous change.

8

ILLUSTRATIVE EXAMPLES FROM WORLD MODELING: ACCEPTANCE, IMPACT, AND CONTINUANCE

Michael Hopkins

When I was last in Washington, I asked a taxi driver the way to the White House. His reply was: "If I were you, I'd go Roosevelt Avenue, turn right at the World Bank, left by Bank of America. No, no, if I were you, I'd go to the Potomac River and go up to the Washington Monument, but when you get that far...no, no, that's not a very good idea either. You know, sir, if I were you, I wouldn't start from here." Often model builders behave like the taxi driver in this story and this, perhaps, is one of the reasons that policy makers have such problems with modelers. The moral of this story is that model builders have to try to understand the constraints under which policy makers work.

If change really comes in a discontinuous way, then there is not much that can be done about it. However, that's adopting a very deterministic view of how the world works and assumes that most discontinuous changes, such as disasters, wars, depressions, and so on, are human-generated. Hence, the way out is to show the consequences of a number of decisions as well as alternatives, rather than to predict a doomsday. Only when well-argued alternatives are presented, backed up with facts, will government and private-sector planners and decision makers sit up and take notice of model builders. Many people have said to me during this meeting that the main customers for modeling in the United States are not necessarily Congress. I agree and believe there is some consensus that modelers have a great deal of work to do in educating people at all levels.

Often I hear the complaint that the kinds of things that modelers produce are difficult for policy makers to understand. I find that rather depressing, because there is no sense in a modeler spending two years analyzing facts, generating models, and at the end of that period preparing only a three-page position paper. The advisors to the policy makers (our first audience) to whom the paper is going are not going to believe the results unless they can actually see the 600- or 700-page report which has been clearly and closely argued--with all the equations and the facts--and the model clearly presented. Even if that report is for only a half-dozen people, I think that's the first report that the modeler should prepare. Then one must go on from there to prepare the three-page position paper or the Washington Post article.

We must ask ourselves: who are the policy makers? Are they interested in our work? Usually they are not very much interested in a thick report because there is much too much information and they don't have time to look at it. However, model builders can become a little bit dangerous to policy makers, because if we sit on a television program with a policy maker who is expounding his latest idea from a very general level, a model maker might say, "No, the facts don't agree with you. I'm very sorry. What specific tax policy were you talking about? What effect will that have on inflation?" Hence politicians in general are not very happy to have a modeler at their side questioning their facts--particularly if the media are involved.

In my own organization, for example, there is concern with such things as population growth, or unemployment, or the effect of wage demands on employment, and so on. Whenever we are discussing a number of issues, certain groups are involved. They say, "Well, there's the population group, there's the technology group, there's the rural development group...and then, there are modelers." They never know quite what to do with the modelers. So far it is difficult to get acceptance of the fact that modeling is just a methodology that can help to analyze population issues, employment questions, or what have you. The modelers might not be able to do it very well, but at least they have a methodology which is generally accepted, which can build on earlier models and so push the frontier of knowledge further forward.

But I don't want to talk much about model philosophy. The issue of mental models, and the role of analytical models, have been discussed, almost ad nauseam, in the last ten to fifteen years. What I want to talk about is the impact, not of modelers on policy makers, but of policy makers and politics on modelers. I should like to give a couple of examples of this from my own experience.

In 1977 the team I was working with completed the first application of the BACHUE model (Bachue is a Colombian goddess of love, fertility, and family harmony). Our work had been financed by the Population Fund at the United Nations (UNFPA), and in 1982 a UNFPA commission (which had had nothing to do with our model) came to town to investigate other issues. I met them at a cocktail party and some of them said, "What are you working on now? Are you working on population?" I said, "At the moment, no, but I did work on the BACHUE model." They said, "Oh, no. How could you work on that? There's been no impact, nothing's resulted from it. Look at the conclusions. You're saying that the anti-natalist theory might not work, that population growth under certain conditions might actually be beneficial. And you said that migration policy might be more important than population policy per se. We just cannot accept that. That was why we decided not to give you any more funds--nothing." I turned around to them and said, "I'm glad you know so much about the project. Tell me about some other project that UNFPA has financed over the last ten years and spent one-quarter million dollars on, that you know so much about." It seems to me that this is one example of policy makers actually understanding the results of the model, but then cutting off funds, because they did not like the results since they were against the political interests of the UNFPA organization.

In the early '80's we extracted some of the technology built into the BACHUE model and created the basic needs models, which we applied to a number of countries, including Colombia. Although no particular policy maker has used the model as such, I still consider our work there successful because, during the process of the project, the model was used as an organizing scheme, and about twenty-five graduate students wrote their theses on aspects of the model. The results of their research served as some of the inputs into the model. These graduate students were very stimulated and understood the model enough not only to use it, but to criticize it--which is no bad thing! All of these students have graduated and gone into industry or government. The person who did the demographic subsystem of the Colombian model (now known as "Colombia 2000") is now head of the census in Colombia. This person has an awareness of models--how they can be used--and also of their limitations. I regard that as somewhat of a success even thought the particular results of the model have not, to date, been widely used.

That is how the work lives on: the technology we developed in the early '70's finds its way into other types of work. That too speaks of success to me. If people were still using the model we developed in 1974 (warts and all), I'd be disappointed. But one thing I've learned, ten years after having developed the BACHUE model, is that the technology lives on and is gradually being replaced with better ideas, so I don't think the earlier experience can quite so easily be written off as useless. I can think of a large number of applied research reports which have been quietly shelved over the past decade--nobody will ever hear of them again.

A third example is the role of the International Monetary Fund. It seems to me that a major force for instability in the next few years may arise from the actions of the IMF as a (sometimes as the) world banker. It is trying to help developing economies in order to put them on the straight and narrow, and thereby help their creditors in the industrialised world. The problem is what one means by "straight and narrow." There is a mathematical model of the IMF, developed in 1957, published in the International Monetary Fund Staff Papers. This model is: devaluation in a developing country having economic problems is coupled with high interest rates, the need to do something about money supply, and always the reduction of public expenditures. Hence the intent is to stimulate savings and investment and to encourage exports.

One problem is that these policies reduce effective demand, an aspect of the IMF model not included back in the '50's. The second problem is: where are all these exports to go? If you're reducing wages all over the world, reducing public expenditure and squeezing economies, who is going to buy these particular exports? One country's exports are another country's imports. I don't believe that the IMF have done enough to study that problem; again, their earlier model is insufficient to do that. This is an example of a particular model having powerful impact because it captures the basic political beliefs of the people using it.

In the ILO, we suggested that the basic needs approach could do a lot to turn some of these economies around. Five years ago, when I did a model of basic needs in Brazil, I predicted the current Brazilian debt crisis. At that time I suggested that if Brazil were to follow another strategy--one which included a better income distribution and the stimulation of the human resource potential--then they wouldn't get into the present deadlock. Unfortunately I could not continue that work in Brazil because they said: "We don't want to worry about basic needs and distribution in this country. This is going to harm our international image--it's not going to allow us to get international financing for our development. So we want to concentrate on the international dimension and we don't want that image to be clouded by this sort of work on basic needs." That is another illustration of a political act affecting a model.

Let me say, then, that model builders are not magicians, are not miracle builders. Basically we're pretty stupid people--that's why we need models, unlike Maynard Keynes, for example, who never used mathematics or models to develop his ideas--he just happened to be extremely brilliant! Nevertheless, Keynes had a mental model and the impact of his model is now well known. Further, that Keynes's model has outlived politicians gives me enough confidence to make a final, controversial point: that a model can eventually be more important than politicians. Who remembers the politicians of that time? Yet nearly everyone remembers Mr. Keynes.

The International Labor Office, where I work, is a specialized agency of the United Nations. It is not that popular in the United States. The Americans, in fact, withdrew from the organization some five years ago, but, happily, have come back since. What I want to discuss now is the experience that we had using a global model.

In Latin America, the Fundacion Bariloche developed a model which had as its core a focus on basic needs. The results of that model were very Utopic; they said that you can meet basic needs for the developing part of the world if you have an equal distribution of income and assets. It was a very Utopic model because it presumed an initial condition of equality in the world. Since we at the ILO were interested in the idea of basic needs and its relation to employment in developing countries, we began to work with the Bariloche group in 1975-76, while at the same time aiming to make the model slightly less Utopic.

We included in the Bariloche model some aspects of income redistribution and we refined the model to cover more regions of the world. Briefly, our results showed that to meet the basic needs of the poorest groups in developing countries within a reasonable time, high rates of growth would be required, or a redistribution of income, or a combination of both. In 1976 the ILO was asked to organize a World Employment Conference. Based on the persuasive arguments coming from this global model, the Technical Secretariat of the ILO were much attracted by the idea of basic needs, and put this forward as the centerpiece of the conference. As a consequence the Ministers of Labor and Planning, trade unions and employer organizations agreed on a world strategy to meet the basic needs of the poorest.
This experience gives an idea of how global modelling can be applied; it was actually used to introduce a concept, a strategy of meeting basic needs for the poorest countries of the world in a certain period of time. However, after 1976 there was a political response to this strategy. First of all, the idea of meeting basic needs was adopted by developing countries in other international fora than the ILO, unanimously; it was then adopted by the industrial countries, the World Bank, the OECD, etc. Shortly after this, in

1977, the strategy of basic needs was rejected by the developing countries, because, they argued, that strategy focussed on the internal problems of developing countries and ignored world problems, such as access for their goods to markets in industrialized countries, coupled with improved terms of trade, more concessionary aid, etc. So developing countries, after agreeing to this idea in 1976, rejected it in 1977, in international circles. Soon afterwards the strategy was also rejected by the developed (or industrialized) countries, which said: "If the developing countries are rejecting it, why, then, are we pursuing it? Aren't we going to get even more unpopular if we pursue it?" So at the international political level, after one year, the basic needs strategy was discarded.

However, at the technical level--at the level of policy advisors, for example--interest remains high. Even to this day in the ILO we get a fair number of requests to advise on the technical aspects of developing further the idea of basic needs. One of the things we have been doing is to develop models at the national level to examine some of the questions that concerned both the developing countries and the developed countries in 1976 and 1977. One aspect we looked at was the relative importance of domestic versus international policies in meeting the basic needs of the poorest people. We then applied a "basic needs model" to a number of countries: Colombia, Brazil, India, Kenya, Sri Lanka.

In examining this aspect, we explored two main international scenarios and two main national scenarios. One of the international scenarios was friendly toward exports of developing countries and was optimistic concerning world trade; the other was the reverse. At the national level one scenario was that the developing countries were export-oriented (assuming a friendly international market), and the other was that they were much more defensive concerning trade and concentrated on developing their own internal human resources and meeting basic needs. Hence, we ended up with four combinations. The results found that in order to meet basic needs the best combination was an export-oriented policy coupled with the rest of the world offering a friendly trading and aid market internationally. Equally powerful was an alternative strategy that unfriendly international markets were assumed and the national economy pursued a basic needs strategy.

What lessons for global modeling can we learn from this experience, in the light of the objectives of this meeting? First, I would say that global models are useful in highlighting particular issues that affect many countries at once. For example, Limits to Growth brought up the issue of population resources in the '70's, the Bariloche model brought up the issue of basic needs, and the UNITAD global model is looking at the effects of growth in the North on employment in the South, and so on. Second, there should not be much detailed research and data collection going into a modelling effort at the global level, because global models only treat issues very, very generally. For example, in the case of basic needs, the Bariloche model could not be used at the national, policy level and new models had to be developed at the country level. Here, if we wish to model adaptive mechanisms through which societies adjust to change, it would be better to develop a simple model first, rather than going to a global model. Third, problems of data and concepts in modeling (e.g., the types of data collected in the Third World) depend on the development model they followed in the past. For example, we can find all sorts of data in the Third World about national accounts, about exports, about growth, about population, and so on, but we can find very few data about basic needs, income distribution, and the allocations of government expenditures to poor people. This is because the development models of the past never emphasized those particular aspects. Fourth, as other speakers have said, to study the future, one must study the past. Hence, discontinuities coming from, say, technical change must be studied to see what happened before. Somebody mentioned that global modelers, in general, do not do enough of this type of hard work. Fifth, if change really comes in a discontinuous way, then there really is very little that we can do about it. To be sure, to accept that means that one is adopting a very deterministic and technocratic way of looking at the world. Most disasters which occur in the world--such as famine, pollution, wars, depressions and so on--are human-generated; they don't come from some planet outside the earth. Hence global models that just predict disaster and doomsday without looking at alternatives are doing a disservice to humankind. It seems to me that only well-argued alternatives, backed up by facts, will cause government and private sector planners and decision makers to sit up and take notice. Unfortunately, however, our impact as researchers is in the medium-to-long-term (ten to fifteen years) range; our impact will never be tomorrow given the short-range concerns of most policy makers.

Coming back to the questions of "Who are the policy makers?" and "Are they interested in our work?" I would say that although we can identify them, they are usually not interested. Perhaps it's because we give them too much information and we become dangerous because we know too much. What politicians do is to resort to their own technical advisors, people they can trust and can challenge. They rarely look to another set of experts and are suspicious of new ideas--that is, if they consult anyone at all, although a good politician would do that. Also, politics itself plays a large role in the way that technical work becomes accepted. For example, a conclusion dictated by a global model to reduce interest rates to stimulate effective demand and hence to stimulate imports and exports internationally, might not achieve acceptance since it would not be politic at a particular moment in a particular country. There might be a sound technical reason to accept such a policy, but a politician won't advance it because he won't be accepted as a leader next time around. Hence it seems to me that our main role as researchers is to make our information and results and facts available to the general public through meetings, speeches, using the media, and so on. If citizens can become more informed, they become more knowledgeable about which politicians they wish to lead them, and those they don't want.

Finally, I want to refer to a point made earlier about models and their relevance to policy makers. Policy makers persist in saying that models often aren't relevant to their decisions; we might put it the other way around and say that quite often models are ahead of policy makers. I think that we should remember one thing: that models are going to be around a lot longer than most politicians!

9

USING MODELS AT THE FEDERAL LEVEL: VIEWS FROM THE CONGRESSIONAL CLEARINGHOUSE ON THE FUTURE

Lena Lupica

The Congressional Clearinghouse on the Future is often perceived as an information broker on Capitol Hill. We broker new and innovative ideas from the general public and from the Congressional support agencies: Office of Technology Assessment, Congressional Research Service, and the General Accounting Office. The support agencies supposedly have a direct line to Members of Congress, but often have a direct line only to those Members who make a specific information request. Because of its ability to initiate work, its capacity to cross Congressional committee lines, and its bipartisan stature, the Clearinghouse could broker to every Member of Congress. Also, by identifying which Member of Congress would be most interested in a piece of information, it could be selective in distributing information. We also choose the medium through which information is exchanged: newsletters, individual meetings with staff and Members of Congress, workshops, or major speakers (often Nobel Prize winners) at Congressional dinner sessions.

A specific example of how information is distributed may be helpful in understanding the Clearinghouse role. Carl Sagan recently met with a few Members of Congress in a personal meeting, in a private office, to discuss "The Day After Nuclear War." The Congressional people were quite impressed by his statements. I had a meeting with Congressman Edgar that day because we were making a presentation about trends to a group of ministers in higher education. Edgar was so excited about Sagan's remarks that he immediately started to educate the group about the possibility of a nuclear winter. Those ministers will return, I am sure, to educate the people in their schools and churches. They will preach the message of a nuclear winter and as a result of our meeting it may go farther than anything else they heard that day.

Charlie Rose, a Congressman from the great State of North Carolina, started the Clearinghouse in 1976. He was in awe of what Alvin Toffler said in 1975 about the rapid changes in our society and thought that Congress needed a foresight structure to provide them with information about the future. The Clearinghouse is one of the more unusual organizations on Capitol Hill because Members of Congress share the services of the staff. I am rarely paid by the Member of Congress for whom I might be doing work that month. The members of the Clearinghouse believe that our foresight work is important enough to share the cost of our salaries. We now have 85 Members of Congress in the Clearinghouse, up from eleven when the organization started.

The Clearinghouse is aided in its foresight activities by two other organizations. We have an advisory group, made up of 25 futurists, which helps guide the Clearinghouse. These advisors help us determine the issues which we concentrate and focus upon each year. For example, the issues for 1984 are the computer/telecommunications merger, alternatives to health care, trade and investment policy and biotechnology. In addition, we have input from the Congressional Institute for the Future, a non-profit and independent organization designed to conduct educational and futures research work for Members of Congress and also to provide that information to the public.

I come to the Workshop on Models and Discontinuities to describe the Clearinghouse and to share my concerns about models and their creators. I am always open to new ideas to broker in Congress, but I am also concerned about the use of models in decision making. I started thinking about models with The Limits to Growth study when I taught a group of high school students what the basic themes of that model were--what they meant for the world--and tried to get them to think seriously about the world and the decisions they made on a day-to-day basis. I carried that interest to my later involvement with students at the university level, urging them to think holistically, systemically. Finally, I was able to reach the continuing education and adult community, particularly in the business sector. Business people have a tendency to buy as many models as they can, because they are unsure which one is really accurate (if any of them are accurate), so they are anxious to compare models. They spend lots of money on models--particularly econometric models. Most econometric models do not include human values, they do not include attitudes. Final decisions are usually based on the underlying values of a person or culture; without consideration of these qualitative factors, models become less effective tools for decision making.

I have now moved into the Congressional scene and I find that the Congress is a very human institution. Members of Congress do things not always based on logic; however, models expect people to behave logically. They don't. So when modelers approach decision makers, there is clearly a need to approach them at a "value" or "emotional" level, not at the "intellectual" level. Carl Sagan approached Members of Congress at the emotional level and consequently succeeded in communicating scientific ideas about a nuclear holocaust.

One of the problems I see in this conference is its élitist quality. There are no minorities here except for two women [sic]. And you certainly have your own modeling language. Most politicians do not understand your language. Élitist groups tend to approach the problem of getting information to Members of Congress in an élitist manner. The communication is to talk to Members in the language of their environment. If you talked to Congressman Rose in North Carolina, he would talk like a real Southerner, but when he is in Congress, he takes on the speech pattern of the Congressional culture. He is one of the brightest human beings I have ever met--extremely knowledgeable on technology, particularly information technology --but some people may not know that in North Carolina.

I am trying to say that my major concern with models is that the information needs to be better and that it needs to be communicated differently. Modelers need to continue working on models and improve upon them. The way to improve them is to continue to observe where the mistakes are and also to add values or qualitative information to the models. Further, people need to believe what the modelers are saying--it doesn't do any good if people do not believe in the forecasts. A lot of missionary work needs to be done here. There are too many forecasts that proved valid--forecasts that some people presented for as many as twenty and thirty years--but failed to have a timely impact because the right individuals were not convinced. Communication of ideas and forecasts in a language that your audience will understand should be primary goals.

In summary, it seems to me that your aim is to influence decision makers, to examine certain forecasts seriously and to be aware of certain direct results from models; the next step should be to act upon them. To accomplish this, you must recognize that decision makers are human--don't approach them totally in an intellectual way, get to know them first, and learn their language. Also, the public should be informed about models and they will help communicate those ideas. Until decision makers, particularly those in Congress, are aware that taking advantage of the insights provided by models is important, and come to believe it is important to their constituents, they are not likely to use information derived from models. There was direct benefit in my using The Limits to Growth model to educate a group of high school and university students, because students, over the years, have become proponents of models and they will have direct impact on policy makers through their voting power.

It is my opinion (and advice) that you need a constituency for the future, and if your élitist approach is not educating the public, then you will be less effective in having your models used. Also, as citizens interested in foresight planning, you should try to elect to public offices those who are sympathetic to such planning; you may even choose to run for office and make a difference.

10

THE SUMMING-UP

Gerald R. Stairs

During this workshop we have explored our conceptual objectives and have surpassed our expectations for agreement. We brought out the notion that point predictions in time are extremely difficult to make; most who talked about the nature of modeling have underscored that conclusion and I believe we all understand from our own experience and agree. Accurate point predictions will, by their very nature, always be difficult to bring about; in terms of the present degree of modeling sophistication we recognize the need to exercise constraint in making such predictions. There was also consensus on the need to strengthen the relationships and communication between policy makers and modelers.

A common denominator which emerged, whether talking from the policy or technical point of view, was that while it may be difficult to provide a point prediction for a future event, what we really seek is to bring about a better understanding of the process, and thus to provide insight to policy making. We can look back at the early writings of Adam Smith, who suggested that the man of systems is always seeking to treat the world like a large chessboard and thus to move the pieces about, failing to recognize that each of those pieces has its own sense of motion and direction. It is obviously not easy to program the world as if it were a giant chessboard; both the policy maker and the modeler must recognize this.

During our discussions several aspects of global modeling emerged that were ancillary to the main purpose of our workshop. These included discussion about data collection, availability, and accuracy, and the utility of models as teaching tools. Issues of data accuracy, continuity, and availability inevitably arise whenever one turns to models as a tool. In relation to our workshop theme the point was made that data collection methods, as well as model assumptions, may be affected by recognition of the potential for discontinuity. Thus one may argue that appropriate attention to historical and currently available data may provide insight into the probability of future climactic events. In discussion of long-wave and short-wave phenomena, coupled with what he referred to as the gentle rain of randomness, Dr. Sterman provided us with a useful view that major catastrophic events may be foreseen with a higher probability than their apparent random nature would suggest. Albeit that such estimates lack precision, they nonetheless serve to encourage a holistic approach to data collection and evaluation, a trait that has high utility over the muddle-through style of policy setting that has been all too prevalent in society.

Several conference participants underscored the pedagogical utility of models. That there is a certain attractiveness to using a model for demonstration of complex issues is a widely-accepted teaching concept, yet it is also a concept that may elicit a great deal of argument about precision and over-confidence or over-simplification if extended too far. An example of both effectiveness and controversy can be found in the first major presentation of world-scale modeling sponsored by the Club of Rome over a decade ago. The World 3 Model and its popular-book version, Limits to Growth, had profound influence in focusing the world's attention upon the interaction between resources, economics, and policy issues. That it failed to satisfy the highest standards of accuracy and predictability does not detract from the generic value gained by focusing global attention on these dynamic issues. The utility as a teaching device was obtained over a wide range of students, from young scholars to major decision makers in private and public sectors. As good teaching should, it also stimulated significant additional inquiry and the development of more refined models, which, in turn, continued the debate about effectiveness on a more evolved level. We recognized in our conference that a major flaw in modeling can be the problem of discontinuity, and at that point came full circle to the realization that models have high utility for forcing attention to the major changes that may alter the course of society on a global scale.

Our sessions provided excellent commentary upon the degree of difficulty that one encounters when moving from qualitative logic to the quantitative and analytic process. The propensity for analytic over-determination was noted, as was the related notion that, while the pursuit of predictability may be difficult, it has the favorable impact of enhancing the development and exposure of relevant data. Several speakers stressed the possibility of applying a theory which in itself is correct, but is one that becomes inappropriate theory in empirical situations. Perhaps this observation is correlate with the common complaint of the lay person: that the theoretical approach violates "common sense" or even direct knowledge of what is actually happening. Again, we came back to the notion of discontinuity as a variant in otherwise logical modeling assumptions.

Several of our colleagues have suggested that discontinuity may be seen as part of the continuum, at least in the sense that some discontinuities may be more easily foreseen than others. These observations returned our attention to the categorical statement on discontinuities. It was pointed out that not all discontinuities come "out of the blue;" sometimes we really do have forewarning but we insist on playing societal brinksmanship at the decision level. Recent examples on the economic scene are the oil embargo of the early '70's and the global economic impact of drastically higher prices; or the U.S. auto manufacturers' response to the need for redefining the size, quality, and fuel efficiency of their products. These comments on the generic level in a theoretical sense, and on the specific level in actual economic terms, highlight both the utility of models and the need for more dialogue between policy makers and modelers.

In relation to discontinuities it sometimes seems that scientists are too cautious about prediction. Major shifts in societal capability or emphasis are often as accurately predicted by the writer of fiction as in the scientific community *per se*. The science fiction writers do seem able to break with conventional continuity and to take bold leaps out into the future with descriptions that lie outside the range of more scientifically considered projections. Paradoxically, these predictions--television, wrist radios, nuclear power, ray guns--have been realized with some consistency as science and technology move forward. Perhaps we do need to consider the allocation of thinking time to some unfettered and discontinuous projections of new discovery, rather than our tendency for linear extrapolation of current technology. While such excursions may bother those who hold close to the classic tenets of scientific inquiry, it would nonetheless provide a means for enhancing consideration of discontinuity in the context of long-term development.

In the final analysis, I believe our session on discontinuity has served most effectively to enlarge our prospectives on global foresight planning. We have noted that an awareness of discontinuous change will affect the design, the utility, and, most importantly, the confidence of utilization of global models. Further, it has underlined a need for effective communication between the policy and decision makers and those who would provide the information base. We have correctly noted that emphasis should be placed upon understanding the process of change, not upon making point predictions. Emphasis upon the latter, and upon displaying the algebraic formulation or algorithms inherent in the modeling world does not necessarily impress or fill with confidence those who see themselves as "real-world" decision makers. There is need for meetings of world-class scholars with world leaders to discuss global foresight. Such awesome models as the one recently developed, presenting the effects of global nuclear war (Sagan and Ehrlich, 1984), along with those under refinement over the past decade, have utility as educational tools for the policy maker.

It is with cautious optimism that I close by saying we are on the right track as educators and scientists when we recommend continued consultation by persons similar to this distinguished group. I do so the end that humanity may be well served by the holistic and ambitious considerations of global foresight planning.

CONTRIBUTORS

Richard Day
Modeling Research Group, University of Southern California

Lindsey Grant
Former Deputy Assistant Secretary of State for Environment and Population Affairs

Michael Hopkins
Employment and Development Department, International Labour Office, Geneva

Barry Hughes
Graduate School for International Affairs, University of Denver

Lionel S. Johns
Office of Technology Assessment, U.S. Congress

Lena Lupica
Director, Congressional Clearinghouse on the Future

Gerald R. Stairs
Vice President for Academic Affairs, California State University, Chico

Robert Smythe
Carolina Resource Consultants

John Sterman
Systems Dynamics Group, Massachusetts Institute of Technology

CONFERENCE DISCUSSANTS

John Caldwell
Chancellor-Emeritus, North Carolina State University

Robert L. Hoffman
University Studies, North Carolina State University

Quentin Lindsay
Science Advisor to Governor James Hunt

Sir James Murray
Former U.K. Ambassador to the U.N. and other international organizations (including GATT) in Geneva, former President of the Security Council